Getting to Grips with BIM

With the UK Government's 2016 BIM threshold approaching, support for small organisations on interpreting, filtering and applying BIM protocols and standards is urgently required. Many small UK construction industry supply chain firms are uncertain about what Level 2 BIM involves and are unsure about taking first steps towards having BIM capability. As digitisation increasingly impacts on work practices, *Getting to Grips with BIM* offers an insight into an industry in change, supplemented by practical guidance on managing the transition towards more widespread and integrated use of digital tools to manage the design, construction and whole life use of buildings.

James Harty lectures at the Copenhagen School of Design & Technology. His PhD at the Robert Gordon University (2012) was entitled *The Impact of Digitalisation on the Management Role of Architectural Technology*. His Masters at University College Dublin (UCD) (1988) compared house typologies against urban morphologies in Dun Laoghaire, while his Bachelor project at UCD (1983) designed maritime facilities at Port Oriel. James started using CAD in 1983 and BIM in 2006.

Tahar Kouider (MCIAT) is Programme Leader of Undergraduate and MSc Design Management programmes at the Robert Gordon University. He trained as an architect then obtained an MPhil from Oxford Brookes University. With 18 years of teaching and research experience, he has published internationally and contributed to the UK BIM Academic Forum and the International Congress of Architectural Technology (ICAT).

Graham Paterson (MCIAT) is an ARB registered architect, chartered architectural technologist and self-builder. Initially trained in mechanical engineering, he has taught technology and product design in the UK and internationally. He is a member of the BIM4SME team tasked by Government to assist SME built environment organisations' engagement with digital technologies in advance of the 2016 BIM mandate.

Getting to Grips with BIM

A guide for small and medium-sized architecture, engineering and construction firms

James Harty, Tahar Kouider and Graham Paterson

Routledge
Taylor & Francis Group

LONDON AND NEW YORK

First published 2016
by Routledge
2 Park Square, Milton Park, Abingdon, Oxon OX14 4RN

and by Routledge
711 Third Avenue, New York, NY 10017

Routledge is an imprint of the Taylor & Francis Group, an informa business

British Library Cataloguing in Publication Data
A catalogue record for this book is available from the British Library

Library of Congress Cataloguing in Publication Data
Names: Harty, James, author. | Kouider, Tahar, author. | Paterson, Graham, author.
Title: Getting to grips with BIM : a guide for small and medium-sized architecture, engineering and construction firms / James Harty, Tahar Kouider and Graham Paterson.
Description: Abingdon, Oxon ; New York, NY : Routledge, [2015] | Includes bibliographical references and index.
Identifiers: LCCN 2015021491| ISBN 9781138843974 (pbk. : alk. paper) | ISBN 9781315730721 (ebook)
 Subjects: LCSH: Building information modeling.
Classification: LCC TH438.13 .H37 2015 | DDC 720.285–dc23
LC record available at http://lccn.loc.gov/2015021491

ISBN: 978-1-138-84397-4 (pbk)
ISBN: 978-1-315-73072-1 (ebk)

Typeset in Bembo
by Out of House Publishing

Contents

Figures

Acronyms

ADT	Autodesk Architectural Desktop
AEC	Architects, Engineers, Contractors
AI	Architects Instruction
AIA	American Institute of Architects
AIM	Asset Information Model
BAA	British Airport Authority
BEP	BIM Execution Plan
BIM	Building Information Modelling
BIS	Business Innovation and Skills
BISDM	Building Information Spatial Data Model
BOQ	Bill of Quantities
BRE	Building Research Establishment
BREEAM	British Research Establishment Energy Assessment Method
BS	British Standard
BSRIA	Building Services Research and Information Association
CAD	Computer Aided Draughting
CAFM	Computer-Aided Facilities Management
CAPEX	capital expenditure
CATIA	Computer Aided Three-Dimensional Interactive Application
CAWS	Common Arrangement of Work Sections
CCPI	Co-ordinating Committee for Project Information
CDE	Common Data Environment
CIAT	Chartered Institute of Architectural Technologists
CIBSE	Chartered Institution of Building Services Engineers
CIC	Construction Industry Council
CITB	Construction Industry Training Board
CLAW	Consortium of Local Authorities in Wales
CMMS	Computerised Maintenance Management Systems
CNC	Computer Numerical Control
CO	Change Order
COBie	Construction Operation Building information exchange
COSHH	Control of Substances Hazardous to Health
CPD	Continuing Professional Development
CPIC	Construction Project Information Committee
DfMA	Design for Manufacture and Assembly
DTI	Department of Trade and Industry
E4BE	Education for the Built Environment

EBD	Evidence Based Design
EIR	Employer's Information Requirements
FM	Facilities Management
GIS	Geographical Information Systems
GSA	General Services Administration
GSL	Government Soft Landings
GVA	Gross Added Value
HEI	Higher Education Institution
HS2	High Speed 2
ICE	Integrated Concurrent Engineering
ICIS	International Construction Information Society
ICT	Information and Communication Technology
IDT	Innovation Diffusion Theory
IFC	International Foundation Classes
IGT	Innovation and Growth Team
IPD	Integrated Project Delivery
IT	Information Technology
IWMS	Integrated Workplace Management of Systems
LOD	Levels of Detail
LOI	Levels of Information
MAC	Marginal Abatement Costs
MPDT	Model Production and Delivery Table
NBS	National Building Specification
NFB	National Federation of Builders
NRM	New Rules of Measurement
nZEB	nearly Zero-Energy Building
OECD	Organisation for Economic Co-operation and Development
OJEU	*Official Journal of European Union*
OPEX	operational expenditure
PAS	Publicly Available Specification
PDT	Product Data Templates
PFI	Private Finance Initiatives
PIM	Project Information Model
PIP	Project Implementation Plan
POE	Post-Occupancy Evaluation
PPP	Public Private Partnerships
QAA	Quality Assurance Agency
QM	Quality Management
RFIs	Requests For Information
RIBA	Royal Institute of British Architects
RICS	Royal Institution of Chartered Surveyors
SaaS	Software as a Service
SAP	Standard Assessment Procedure
SCC	Social Cost of Carbon
SME	Small and Medium-Sized Enterprise
SMM	Standard Method of Measurement
SPC	Special Purpose Company
TAM	Technology Acceptance Modelling
Uniclass	Unified Classification for the Construction Industry

Introduction

A central lesson of science is that to understand complex issues (or even simple ones), we must try to free our minds of dogma and to guarantee the freedom to publish, to contradict, and to experiment.

Carl Sagan (1998)

According to Government figures, construction contributes almost £90 billion gross value added (GVA) to the UK economy, representing almost 6.7 per cent of total output and sustaining 280,000 companies. In 2006, construction's GVA represented twice that produced by the energy, automotive and aerospace sectors combined. The built environment supports around 3 million jobs, equivalent to 10 per cent of total UK employment, and has been estimated to contribute 70 per cent towards UK manufactured wealth. The sector is key to the Government's long-term objective of making the UK a low-carbon society, with buildings accounting for around half of greenhouse gas emissions.

It has been argued that construction has a deep significance in relation to the development and well-being of the UK socio-economic fabric. Construction creates, builds and maintains workplaces which enable businesses to operate and grow, as well as the homes, schools, healthcare, leisure and related infrastructures which support and enhance contemporary life. Many of the construction projects which populate our cities, towns and landscapes are executed by large contractors, some part of multinational conglomerates. However, even the largest UK construction company only holds a 3.5 per cent share of the market, and around 90 per cent of the UK construction supply chain comprises the small and medium-sized enterprises (SMEs) and micros described by the European Commission as 'the engine room of the European economy' (EC 2013).

UK construction is diverse in nature and fragmented both by international standards and in comparison with other domestic sectors. The built environment marketplace is highly competitive and profit margins tend to be low (typically between 2 and 3 per cent). While statistics vary year-on-year, construction supports more than 270,000 active enterprises. Over 90 per cent of the 186,000 companies in construction contracting employ fewer than 10 workers, and around 70,000 businesses are one-person operations. Professional services provision is similarly diffuse, with around 23,500 companies employing 225,000 people. Within the wide-ranging spectrum of built environment-related activities, many business models may apply. From any angle, it is difficult to envisage one-size-fits-all solutions to the many challenges facing the built environment supply chain in reaching out towards attaining Government targets for a digitally smart, technologically advanced and sustainable industry by 2025.

Even designing and constructing a small building can involve significant levels of organisation, dialogue and information sharing within the supply chain. Laugier's primitive hut, an eighteenth-century literary allegory, defined basic construction elements – post, beam and pitched roof – as basic necessities to modify natural environments and provide human shelter. Imagine the hut fast-forwarded into 2015. To realise Laugier's theoretical concept would conceivably require interventions by designers, engineers, cost consultants, project planners, subcontractors and suppliers. Beyond the basic design and construction tasks, there are probably others feeding into the consultative and statutory processes, which are incumbent on contemporary processes for making buildings.

Today's post-industrial society, characterised by divisions of labour prevalent since the Industrial Revolution, has developed many new mutations which continue to subtly modify historical roles and responsibilities. In construction, for example, it is unlikely that the practitioner who advertised services as 'architect and surveyor' at the end of the nineteenth century would be offering a comparable portfolio of services in today's risk-averse marketplace.

In the main, built environment education is pitched towards undergraduates entering the industry and practising as generalists fulfilling discipline-specific roles. Identities associated with these models for practice are reflected in the ways that construction professionals are educated and trained (for example, in the structure and nature of the educational process, syllabus content and learning outcomes). Also the self-identities maintained by the bodies responsible for stewardship and governance of the various UK built environment professions may influence mutual interactions, particularly when new ways of working enter the frame.

Teamworking principles may be taught during the undergraduate experience. Sometimes construction disciplines in education and training collaborate in simulating real world paradigms for making buildings, but these instances are rarely the norm. In the workplace, the dependence of many organisations on information technology to manage business processes and workflows has generated specialist roles, such as BIM and information managers, to organise and feed increasingly complex data sets into design, construction and post-occupancy building management processes.

Technology has been interpreted as the manifestation of our constant endeavour to improve and enhance the human experience. In construction, that spectrum of activity can range from small functional structures to large and complex buildings. Basic tenets of functionality may be overlaid with (or in some cases subsumed by) aesthetic intent. Historically, drawings have fulfilled such a significant role as communication media in building design, construction and maintenance that it is difficult to envisage life without them. The tradition of abstracting clients' programmatic requirements and professional advisers' design intent into orthographic views set out on paper has endured for hundreds of years and continues to underpin architectural education. As Bruce Lindsey noted in his analysis of Frank Gehry's work methods and digital tools, 'design tends to be a habitual practice'.

The shift from hand to Computer Aided Draughting (CAD) drafting from the 1980s represented an incremental rather than step change in work practices. The means of inputting brain-to-hand information morphed from manual draughting to digital media, but the output remained substantially the same: primarily paper drawings. The facility which electronic drawings offered to layer information meant that the way data was structured within digital files changed significantly. New hierarchies and conventions were established. The role of the CAD or Information Technology (IT) manager evolved to structure and manage digital data within organisations and across design teams. In small companies, sole practitioners and micro mixed-discipline organisations, the design principal and IT manager roles might have been fulfilled by the same person.

The number and configuration of drawings necessary to make a building have also changed and grown over the years; some might argue that growth has been exponential. In the eighteenth century, artistic conceptual sketches could be interpreted, developed and realised by a master builder and a team of craftsmen and artisans. Burgeoning industrial expansion during the second half of the 1800s was serviced by workers often living in tenement dwellings and the brick terraced housing which characterised much of England's urban grain.

In urban conurbations like Glasgow, mass housing was built from minimal drawn information, often just a single sheet of orthographic line drawings to cover architectural design, structure and services. Developer builders assumed the risks for design and construction. Towards the end of the nineteenth century, as rule-of-thumb methods for structural design became replaced by calculation and subject to statutory control, so the number of drawings increased and their nature changed. Technical content became more significant, and the layering of working drawings by scale into general arrangement, assembly and detail continued as a convention throughout the twentieth century.

As more drawings and written specifications were produced for projects, so the potential for inconsistencies and errors increased. As Mervyn Richards (2010) has written, evidence suggested that improving the quality of production information reduced both the cost of developing contract documentation and the incidence of site problems leading to cost overruns and delays. In 1979, the Co-ordinating Committee for Project Information (CCPI) was established by the Royal Institute of British Architects (RIBA), Royal Institution of Chartered Surveyors (RICS) and others with the key objective of bringing about an improvement in the documentation used for the procurement and construction of buildings. The efforts of the CCPI (later Construction Project Information Committee (CPIC)) were directed towards establishing protocols for consistent structuring and management of information across multidisciplinary design and construction teams. The focus for improvement was on both the technical content of the documents and the effectiveness of the co-ordination between them.

During the late 1980s, the launching of the Heathrow Terminal 5 (T5) development marked a further watershed in the generation and management of construction project information. Bob Garrett (2003) of Exitech Ltd, which consulted on the project, noted that T5 actually bundled together 18 major projects. British Airports Authority (BAA), as process-smart commissioning clients and developers, were resolute that the development would be delivered on-time and on-budget. From a design management perspective, the most significant technological advance was the proposition that a Common Data Environment (CDE) would be used as a shared container for the digital data necessary to sustain design, construction and whole life maintenance. Additionally the project incorporated building elements as 3D virtual objects, which combined geometric attributes with metadata embedded as property sets for each object. The subsequent Avanti project, supported by the UK Government through the Department of Trade and Industry (DTI), sought to achieve significant gains in project and business performance via the deployment of digital technologies to support collaborative working through the use of consistent information and design process management protocols.

Time frames necessary to realise building projects can range from months to years. Sometimes things run smoothly, other times less so. Challenges, such as programmes overrunning and costs spiralling out of control, can test clients' faith in their professional advisers to the limit and painfully raise awareness that procuring a building is not comparable with ordering a machine-manufactured artefact like a vehicle, electrical appliance or other factory-produced commodity. Very often insignificant communication difficulties can become compounded into major issues, which can push expectations for a building's realisation way out of shape. In that broad context, it could be argued that there is not a great deal of point in continuing to use manufacturing as a template for a brave new world of automated and highly efficient building production. Since the 1990s, successive UK industry leaders have resolutely tried and failed to apply industrial analogies to a diverse and fragmented sector characterised by a proliferation of small and medium-sized organisations. The analogies are just not valid in many, if not most, situations.

In that context, over a quarter of a century has elapsed since the Boeing Company made the strategic decision to migrate to a paperless environment for aircraft design and production for the 777 generation of passenger aircraft. Manufacture of the previous 747 model had involved around 75,000 drawings to articulate technical specifications. Using information technology for teamworking to link design with production was a key driver. Similarly, whole life considerations, including maintenance in service, were embodied in the frame of reference. These factors energised the subsequent process models developed and adopted by the partnership between Boeing, IBM and Dassault Systems, signed in 1988.

Perhaps what some might consider a rather obvious premise – having timely and immediate access to shared information necessary to develop efficient workflows – was in the frame as well. Whatever analysis is applied in the round, the decisive paradigm shift from paper drawings to digital media was truly a game changer. With notable exceptions, including the formation of Gehry Technologies as a stand-alone company in 2002 and the groundbreaking T5 project, mainstream construction practice (in particular that of small organisations) has just not realised the full potential offered by digital technologies. The idea of digital data flowing seamlessly within design and construction teams between different software applications and across project timelines is some way removed from reality at this point in time.

The UK Government's 2016 threshold for Level 2 BIM compliance may force some built environment organisations to examine how they handle and share digital information as part of the process of making buildings and infrastructure. Given the UK's fragmented and multifaceted construction supply chain, it is conceivable even the smallest organisation, for example a specialist supplier, could be tendering as a partner for a publicly funded project. But, as Woudhuysen and Abley (2004) has argued, the construction sector has a demonstrable track record of taking a long time to fall into line with technological step change. In playing catch-up with manufacturing, much of the built environment's morphology still exudes cottage industry characteristics.

Technology is a fast-flowing river: the pace of change is relentless and all pervasive. The impact of geo-systems and cloud computing is impacting on business activities and daily lives alike. Digital data is being grabbed, filtered down and presented to users more immediately than ever before. Information-rich and geo-smart technologies are at the cusp of being embedded into workflows for manufacturing, while the built environment, with some notable exceptions, is still getting to grips with technological concepts and practices which have been in existence for many years.

Given the structure and track record of our 'industry', it is not surprising that in today's so-called digital age, embedding new and emerging information and communication technologies (ICTs) into built environment culture and work practices has proved challenging, particularly for small organisations and educators. The sheer volume of references to BIM in design and construction media may be pushing some built environment organisations towards examining how they generate, manipulate and share digital information. Others may be constrained by the belief that BIM does not fit their business models or that it is just not for them, or by an opposition to industrial change and innovation.

Over the last few years, BIM language, conventions and protocols have been propagated relentlessly through online communication tools – social media and the like. The BIM landscape and its constituent parts have been charted out, blogged about and discussed across professional bodies and industry organisations. BIM has provided a conduit to sell software and develop new digital tools. The BIM mantra has become diffuse internationally as a catalyst for efficiency gains, cost savings and carbon reduction.

One perceived effect of globalisation is the practical difficulty of separating local, regional and international agendas across many fronts (social, economic, technological and the like). BIM is no exception. BIM conventions and standards are being set and applied both locally and globally. Not unlike the green agenda which percolated through built environment thinking and practice during the 1990s, mixed messages can overlap, become blurred and, in some cases, inhibit decision-making. Greenwash has been supplanted by BIM-wash.

Driven by a transformational vision which embodied digitisation and the use of advanced materials and new technologies, the UK Government's *Construction 2025* (Department for Business Innovation & Skills (BIS) 2013) strategy document set ambitious performance targets for the construction sector, including a 50 per cent reduction of greenhouse gas emissions. In 2010, the Government's industry-wide consultative Innovation and Growth Team (IGT) (Anon 2010) reported on the means by which the UK construction industry could rise to the challenge of the low-carbon agenda. The IGT study envisaged a future, beyond 2020, in which holistically designed, low-carbon sustainable buildings and structures were the norm. This new world anticipated better control of waste, increased use of passive techniques which minimise depletion of resources, and more prefabrication of buildings and infrastructure.

The IGT argued that the key precursor to delivering on performance targets was that Government, the industry and the institutions needed to adapt and secure real behavioural change in their own operational activities. Change could only be effected through a bottom-up approach to multidisciplinary education, with undergraduates and apprentices rising right up through each organisation and the industry. The IGT believed it was necessary for this process to begin immediately. Many construction sector organisations have called for increased pan-industry collaboration to provide a unified sense of purpose for the sector.

The Government's commitment to reduce carbon and other greenhouse gas emissions is now a matter of legal obligation. The strategy by which this target might be achieved will reach deep into every aspect of the built environment, and depends for its delivery upon the construction industry pulling together. Over the next 40 years, the transition to a low-carbon economy can almost be read as a business plan for construction. For built environment organisations, delivering on the task in hand was deemed to break down into three key strands:

- for companies to decarbonise their own business activities;
- to provide people with buildings that enable them to lead more energy efficient lives; and
- to provide the infrastructure which enables the supply of clean energy and sustainable practices in other areas of the economy.

IGT concluded that innovation would be required to stimulate new ways of working and the acquisition of knowledge and skills to provide competitive advantage at home and internationally and, in a highly competitive global market, to consolidate the UK's reputation as a world leader in sustainable design.

BIM is a prime mover in the process of making these aspirations realisable. Predictive energy analysis can be carried out with increasing accuracy at each stage of the design process. Post-occupancy analytical techniques using digital technologies can validate performance in use or pinpoint where shortfalls in predicted energy performance in use are occurring. Daniel Doran of the Building Research Establishment (BRE) argued that it was now viable to produce software tools which offered automated building level assessments of energy performance and carbon loads. With that automation came the processing power for improved functionality, greater accuracy, detailed breakdown of results, and compliance with performance-driven standards. BIM's capacity to instantly extract geometries and metadata from digital models means that tasks previously carried out manually can now produce analytical and summative data in seconds.

It is approaching 15 years since the vast Heathrow T5 programme of works put down a marker for the use of digital media to drive built environment teamworking. For many large construction organisations collaborative BIM now represents a norm. Yet, perhaps not unsurprisingly, the BIM paradigm has so far failed to disseminate downstream to what the writer Chris Anderson (2009) described in a business context as the 'long tail' of SMEs and micros in the UK. Why should that be? Are small construction industry organisations inherently resistant to change? Do they operate different business models to large organisations? Have the benefits which BIM can bring to enhance organisational efficiency been overhyped? Or is there some other reason which has not been identified and articulated so far?

Despite BIM's new and sometimes arcane language as it is used to describe familiar built environment concepts, has anything really changed? Clients still require buildings which match their aspirations and functional requirements to be delivered on programme, within budget and to the specified quality standards. Ease of access to the data sets necessary to facilitate building design, operation, maintenance and future adaption is a given and not some new requirement which has entered the frame of reference with BIM. Evolving processes for information management can simply facilitate access to data. BIM does not de facto produce better buildings. BIM is viewed by the authors as a means which can be deployed to facilitate process. Ultimately it is up to commissioned design and construction teams to agree with their clients whatever arrangements will best enable the achievement of project goals and outcomes.

The BIM2050 Group (BIM2050), comprising 18 young construction professionals from all areas of the industry, was formed in September 2012 by the Construction Industry Council (CIC) to work in partnership with the HM Government UK BIM Task Group and look ahead in developing a vision for the future of UK construction. BIM2050 noted in its 2014 report (BIM2050 Group, 2014) that the UK construction industry's future was faced with high levels of complexity, extreme competition and uncertainty. These issues were related to the outcomes of climate change and resource availability, compounded by the unpredictable and disruptive nature of innovation. Regardless of their origin, it was predicted that these factors would stimulate construction sector change, which would directly impact on every aspect of the built environment.

As a snapshot of building futures, BIM2050 argued that while the 'hard skills', such as engineering, would remain a core construction activity, it appeared that the sector's reputation for inefficiency prevailed, compounded by a lack of collaborative soft skills and poor cultural integration between education and the workplace. In that context, the technological challenges presented by BIM, its antecedents and progress from Level 2 to Level 3 measures of maturity tended to be subsumed into a much bigger and more significant industry picture, which was defined by cultural rather than technological challenges.

As built environment practitioners, educators and researchers, the authors have engaged with digital media for design and construction since the 1990s. As all contributors come from design backgrounds, that ancestry may manifest itself in the tone and emphasis of the book. Having made the point, the writers all have a wealth of experience of interdisciplinary collaboration in the field, as design managers and in teaching future construction professionals from a broad range of discipline backgrounds.

The overarching intention of this book is to offer a practical and grounded approach to assist small organisations and educators in getting to grips with key principles and practical applications for collaborative working using digital tools. It answers such questions as: what steps are necessary to set up and document an organisation's first BIM project, and do BIM standards, such as those set out in the PAS 1192 suite of documents, have to be followed to the letter, or is there scope for adaption and modification to suit project-specific circumstances? Having said that, the authors are resolute that BIM can represent a way of doing things, but may not represent the most appropriate methodology in all situations. The BIM acronym is often used loosely and that is not necessarily problematic. Although the PAS 1192 suite of documents sets up protocols, these may be interpreted differently in different situations by design and construction teams. Viewed through the Government's Level 2 lens, BIM hybrids may evolve which align with the core principles, but develop a range of unique characteristics.

By following a pathway which focuses on basic principles and first steps, there is total commitment to the idea that BIM is not just applicable to the delivery of large and complex building and infrastructure developments. The authors maintain that the structured and efficient use of digital data is equally valid for developing small projects and will argue that case consistently throughout the book.

The authors have also attempted to develop a grounded and consistent thematic schema for the text which:

- takes into account situations where small organisations may be engaging with BIM from a limited resource base. For example, it is unlikely that a sole practitioner would have the need or means to employ a BIM manager with specialist knowledge;
- recognises that adding value, monetary or otherwise, would constitute a clear business incentive to SMEs and micros to adopt BIM;
- investigates and records a range of positive outcomes which can be achieved using low-cost and ubiquitous digital tools;
- in offering a step-by-step approach, acknowledges the reality that migrating from lonely to collaborative BIM may involve significant change in work practices;
- uses language, terminology and process models, such as the RIBA Plan of Work 2013, which small design and construction-related organisations will already be familiar with;
- capitalises on the potential of SMEs and micros as lean and agile organisations to adapt to change quickly and efficiently;
- encourages the generation, use and management of digital data in structured and efficient ways;
- demonstrates that BIM is both feasible for small organisations and can add value to business development;
- promotes lateral thinking in discovering and capitalising on new ways of doing things, even in competitive business environments; and
- advocates keeping a vigilant eye on the potential of new and smart technologies to offer a competitive edge.

The book has been structured in four parts. Parts I, II and III are each divided into four sections. Part IV is divided into six sections. While the book flows from Parts I to IV, the Parts are not necessarily interdependent. For example, an undergraduate student might find it help-ful to refer to Part I during the process of developing a literature review for a dissertation. A practitioner might find it more useful to focus on the practical guidance covered in Part IV. The content of each Part is summarised as follows:

Part I: the BIM journey

Part I analyses the timeline over which BIM has developed and examines a range of drivers for change, including reviewing the roles of Government, professional bodies and industry organisations. Topics discussed include the way in which buildings are made and used, discip-line identities, interrelationships, the evolution of co-ordinated project information and the changing nature of information management. International, regional and national perspectives on BIM are examined in the context of globalisation and what has been described as the 'third industrial revolution'. Early BIM adopters are identified and reviewed, particularly in relation to technology transfer from other industries.

Part II: a contemporary view of BIM in the UK

Part II presents inward- and outward-looking snapshots of BIM from a UK perspective. It examines a range of drivers for change and innovation, including international and national BIM protocol paradigms for data modelling and management in relation to construction procurement. Issues facing educators and professional practice are flagged up, investigated and reviewed from educational and professional practice perspectives. Challenges, risks and potential benefits for SMEs and micros in facing up to the necessity to engage with digital tools in a competitive and volatile marketplace are discussed, including tailoring BIM to suit business models, and filtering out achievable BIM outcomes from generic and bespoke aspects of practice.

Part III: teamworking and information management

The focus with Part III is on setting up and managing teams engaging with BIM scenarios, including the role of clients. Part III addresses a range of paradigms, including lonely BIM and collaborative working. The significance of taking a whole life view with BIM is investigated, including embedding soft landings principles into project planning and realisation. Procedures for setting up and managing CDEs are identified and the value of achieving smooth infor-mation flow is addressed. The role of the BIM/information manager is discussed, including the key task of integrating workflows to ensure that client-driven whole life objectives and outcomes are realised. Thematic aspects of BIM futures are outlined, including big data, the Internet of things and interactions with augmented reality environments.

Part IV: setting up and running a BIM project

The final Part of the book covers a range of issues associated with setting up and running a BIM project. This Part is informed by exemplars appropriate to the authors' intention to focus on the needs of small organisations.

Section 13 identifies and reviews organisational structures and business models which characterise SMEs and micros. Practical aspects of data sharing and exchange are discussed along with protocols for file management within and across organisations. Ethical, legal and professional liabilities are discussed. Specific reference is made to BS1192:2007, or as updated, as a key organisational reference document.

Strategic aspects of applying the PAS 1192 suite of documents to projects are reviewed in Section 14. While the book makes reference to all component parts of the PAS suite, the focus is on PAS 1192-2 as the key instrument for defining and articulating roles and responsibilities, defining and populating CDEs, data-in and data-out, plus the definition and use of embedded organisational standards.

Section 15 reviews the application of appropriate process models tools which shape and track project development over time, in particular, use of the RIBA Plan of Work 2013 and RICS NRM 1,2,3. Communication within design teams, and strategies for cost planning and control are identified along with variations on the generic model suggested by PAS 1192-2 to fit a range of procurement typologies.

As a project moves through the design phase and towards construction and use, Section 16 reviews various shades of BIM and focuses on inputs and outputs in relation to collaboration and workflows, in particular populating information exchanges and defining appropriate levels of definition for model development at each stage of the Plan of Work. Industry classification systems are also identified and discussed.

Section 17 reviews the post-occupancy phase including soft landings and the Building Services Research and Information Association (BSRIA) five-stage process, PAS 1192-3 and the fit of post-occupancy considerations with information flow during design development. In addition, embedding information management strategies for planned and reactive maintenance and incorporating the user experience into post-occupancy evaluation.

Section 18 examines how project performance in use and feedback can be incorporated into organisational structures; using and maintaining the BIM inventory as an 'as-built' database; reviewing how BIM can be capitalised on as a medium for cross-disciplinary partnering to maintain a competitive edge and learning from data management experiences to inform future project development.

Whether practitioners, academics, students or others with an interest in BIM, all readers are encouraged to use the book to interact with their own experiences and project work. Drawing from the literature and industry exemplars, the book is intended to develop a commentary which sets out broad principles for BIM, with an emphasis on the needs and aspirations of small organisations (the SMEs and micro-SMEs). It is not the authors' intention that the book is used to inform the learning and use of particular software packages. Where instances of proprietary software use have been cited in examples and illustrations, no endorsement is offered or implied by the authors.

Ultimately BIM is about how we use and manage digital data to attain practical goals and useful outcomes, be they shaped by clients industry visions or simply the need to achieve more sustainable national, regional and global built environments.

Bibliography

Anderson, C. (2009) *The Long Tail*. Random House Business Books.

Anon (2008) *Construction Matters*, report (Vol. 1) of House of Commons Business and Enterprise Committee, The Stationery Office Ltd, [online]. www.publications.parliament.uk/pa/cm200708/cmselect/cmberr/127/127i.pdf [accessed 8 December 2014].

Anon (2010) *Low Carbon Construction Innovation and Growth Team, Final Report*. HM Government, [online]. https://www.gov.uk/government/uploads/system/uploads/attachment_data/file/31773/10-1266-low-carbon-construction-IGT-final-report.pdf [accessed 6 March 2015].

Anon (2014b) 'Frank Gehry's Technology Company Bought by SketchUp Owners', *De Zeen Magazine*, 8 September, [online]. www.dezeen.com/2014/09/08/trimble-buys-frank-gehry-technologies/ [accessed 8 December 2014].

Anon (2015) *Constructing a Better Future: Achieving Quality and Best Value in the Built Environment*, Construction Industry Council, [online]. http://cic.org.uk/news/article.php?s=2015-03-17-cic-election-briefing-published-today [accessed 24 March 2015].

BIM2050 Group (2014a) *Built Environment 2050: A Report on Our Digital Future*, London: Construction Industry Council, [online]. www.bimtaskgroup.org/bim2050-group/ [accessed 24 March 2015].

CCPI (1987) *Co-ordinated Project Information*, Co-ordinating Committee for Project Information.

Department for Business Innovation & Skills (2013) *Construction 2025: Industrial Strategy for Construction – Government and Industry in Partnership*, London: HMSO, [online]. https://www.gov.uk/government/publications/construction-2025-strategy [accessed 8 December 2014].

Doran, D. (2013) 'BIM and the Eco Footprint', *Architectural Technology*, Issue 107, CIAT, [online]. www.ciat.org.uk/en/other/document_summary.cfm/docid/173321C0-E00A-4BBE-B7494337387030AF [accessed 24 March 2015].

EC (2013) *What is an SME?*, European Commission, Enterprise and Industry, [online]. http://ec.europa.eu/enterprise/policies/sme/facts-figures-analysis/sme-definition/ [accessed 19 October 2014].

Garrett, R. (2008) 'Heathrow Terminal 5: Take-Off for Design Data Modelling', *Design Productivity Journal*, 4(4), [online]. www.excitech.co.uk/dpj/AEC/Vol4_4_pdf/T5_Legacy.pdf [accessed 8 December 2014].

Hood, N. (2013) 'Top 100 Construction Companies 2013', *The Construction Index*, [online]. www.the-constructionindex.co.uk/market-data/top-100-construction-companies/2013 [accessed 8 December 2014].

Hvattum, M. (2004) *Gottfried Semper and the Problem of Historicism*. Cambridge University Press, [online]. http://catdir.loc.gov/catdir/samples/cam041/2003046083.pdf [accessed 8 December 2014].

Lindsey, B. (2001) *Digital Gehry: Material Resistance/Digital Construction*. Basel: Birkhauser – Publishers for Architecture.

Markille, P. (2012) 'A Third Industrial Revolution', *The Economist*, 21 April, [online]. www.economist.com/node/21552901 [accessed 8 December 2014].

NIST (2011) *The NIST Definition of Cloud Computing*, National Institute of Standards and Technology, US Department of Commerce, Special Publication 800-145, [online]. http://faculty.winthrop.edu/domanm/csci411/Handouts/NIST.pdf [accessed 8 December 2014].

ONS (2014) *Construction Statistics*, No. 15, 2014 Edition, Office of National Statistics, [online]. www.ons.gov.uk/ons/rel/construction/construction-statistics/no–15–2014-edition/art-construction-statistics-annual–2014.html [accessed 8 December 2014].

Petroski, H. (1997) *Invention by Design: How Engineers Get From Thought to Thing*. Cambridge, MA: Harvard University Press.

Richards, M. (2010) *Building Information Management: A Standard Framework and Guide to BS1192*, BSI.

Sagan, C. (1998) *Billions and Billions: Thoughts on Life and Death at the Brink of the Millennium*. New York: Ballantine Books.

Woudhuysen, J. and Abley, I. (2004) *Why is Construction So Backward?* London: Wiley-Academy.

Part I
The BIM journey

1 Historical paradigms

Figure 1.1 PhD facility project

1.1 Making and using buildings, people and processes

Thom Mayne said during the *Building Information Panel Discussion* (Mayne 2006) at the American Institute of Architects (AIA) National Convention in Las Vegas that: 'It's about survival. If you want to survive, you're going to change; if you don't you're going to perish. It's as simple as that … you will not practice architecture, if you are not up to speed with this…'. This was in response to: 'By what means do the architects in this audience accelerate their understanding of this new technology and all its implications for practice?' During the discussion, Patrick MacLeamy added: 'This is happening, get with it or get over it. If we don't do this, I don't believe that we are going to be in business.' Equally compelling and from the complete opposite end of the scale, a study in Hong Kong by Tse *et al.* (2005) saw the barriers around this time to BIM adoption as:

- no need to produce BIM;
- existing CAD systems were adequate;
- no desire to commit to extra cost;
- lack of skills;
- BIM could not reduce drafting time;
- not enough features;
- not required by clients; and
- not required by other project team members.

Of these, arguably the only remaining barrier is lack of skills. At that time, BIM was seen as a technology. Now it is accepted as a process (Eckblad *et al.* 2007).

Above and beyond the technology and the management, there is also BIM's impact on society, in relation to how buildings shape our lives and influence our quality of life. Barrett addresses these concerns in looking at both what society deserves and how new methodologies augment change (Barrett 2011). Underwood and Isikdag also reflects on the impact of BIM where it is often seen by management as a utility tool (and not an intrinsic tool), meaning that it has been strongly affected by the economic crisis of 2008, where there have been significant cutbacks (Underwood and Isikdag 2011). This too has a large bearing on society and the quality of buildings produced. It does not go unnoticed that BIM offers worthwhile potential for growth, if properly adopted.

But in relation to this book, how BIM affects management regarding its implementation, sees four emerging dimensions:

- an integrated environment;
- distributed information;
- up-to-date information; and
- new derivatives of the information.

Underwood concluded that relatively new information technologies, such as cloud computing and sensor networks, could readily accommodate these societal needs.

To achieve the full potential of the new technologies, it is necessary to understand project information flows across all stakeholders. For example, speciality trade contractors at the end of the supply chain have noted a 67 per cent improvement to the flow of project information in the last two years (Bernstein and Russo 2013).

1.2 Discipline identities and interrelationships

One of the key disciplines in the construction discipline mix is the technologist. Professionally speaking, it could be said that architectural technologists were spawned out of the modern movement after the Second World War, amid a transition where the focus moved increasingly into the studio or drawing office. This represented a change in the building sector from crafts and handwork to technique, and from apprenticeships to documented processes and methods. This change can be charted right through the modern movement, where it relied heavily on

Figure 1.2 PhD facility project – cross section

new performance requirements, innovations and a much broader use of materials. Essentially, what can be chronicled is that responsibility and decision-making moved indoors.

Technology, and, more importantly, architectural technology are very relevant today. A technologist is one who can implement, understand and communicate the new requirements in a salient, appropriate manner, stressing who is the bearer of this news and who can guarantee it safe passage, to those who can best be served by better-informed design, whether they are clients, users or society at large. Sustainability, energy consumption and embodied carbon are all a result of poor management and abuse of the finite resources this planet has to offer.

Construction activities, in general, are embedded in traditional methods, and many modern technologies are still derived from traditional knowledge or know-how, as is easily demonstrated in masonry, carpentry and plumbing. But there are also new drivers for innovation, for improvements in productivity, including performance, health and safety, quality management and a building's unique qualities. The latter include sustainability, life-cycle analysis, purpose-in-use and the need for the present generation not to compromise future generations' right to the same concession, meaning that change must be implemented, and new or better ways of procurement must be deployed to uphold that right:

> Any technology is but one of many systems that presuppose a building: ecological systems, economic systems, social systems, etc. These multiple systems can be summarised by what Lewis Mumford called 'technics,' a term that describes the relationship between the technical systems and social systems in large-scale habits of mind and action in civilisation. Technics illustrates the collective theories, techniques and technologies that characterise paradigms of choices and practices.
>
> (Moe and Smith 2012: 1)

Le Corbusier, in *Towards a New Architecture* (1921), noted that construction is at a designer's fingertips in the same way that grammar is there for a thinker. But with an ever-increasing churn bringing continual change, this bank of experience is even harder to amass and accumulate. Mies van der Rohe distilled it even further:

> Architecture depends on its time. It is the crystallization of its inner structure, the slow unfolding of its form. That is the reason why technology and architecture are so closely related. Our real hope is that they grow together, that someday the one will be the expression of the other. Only then will we have an architecture worthy of its name: Architecture as a true symbol of our time.
>
> (Mies van der Rohe 2012)

Stephen Emmitt drew a comparison much less with time than with the society in which it is found:

> If architecture is concerned with making society, it is the materials, components and fixings – the architectural technologies applied to abstract ideas and concepts – that helps to realise the built fabric in and around which society functions.
>
> (Emmitt 2012: xii)

He also drew attention to a comment the technicians' society made back in 1984:

> 'Architectural technology is the constructive link between the abstract and the artefact.' CIAT (formerly SAAT, 1984)
>
> (Emmitt 2012: xii)

With these quotes we get a wonderfully compelling notion of what architectural technology can aspire to: becoming the expression of architecture, the realisation of our environment and the link forming the artefact. Padding this out, it can be said to affect the times we live in and the society of which we are a part, while breathing life into the abstractions of designers.

The synergy that drives the architectural technologist could also be said to derive from the artistic (which encompasses design and is somewhat intangible, albeit creative), the procedural (which is deliverable and managed) and the practical (which is functional, tangible and technological) (Emmitt). The cornerstones of the design manufacturing process start with a conceptual cycle where ideas are hatched. This evolves into a functional phase where there is an assessment of fit-for-purpose, to an aesthetical stage where the form emerges. From here, there is a transition from design to manufacture called fabrication. Finally, there is an underlying process to underpin the methodologies used in all the above operations (Kowalski 2012).

The Tower of Babel analogy has left a legacy that has never properly been addressed. Whatever about reaching the heavens and getting above our station, I could never resolve why the good Lord would wish to derail collaboration and introduce counterproductive devices into common parlance. Furthermore, the punishment for building violations by the Babylonian King Hammurabi's building codes 4,000 years ago, stated:

> 229. If a builder has built a house for a man, and has not made his work sound, and the house he built has fallen, and caused the death of its owner, that builder shall be put to death.
> 230. If it is the owner's son that is killed, the builder's son shall be put to death.

Medieval builders knew a great deal about how to build wonderful cathedrals, but their answer as to why those buildings stood up was to point to the hand of God, with a nodding testament that they followed traditional rules and mysteries of their craft.

Perhaps because they were too much obsessed by the moral significance of good work-manship, the old builders and the carpenters and shipwrights never seem to have thought at all, in any scientific sense, about why a structure is able to carry a load. … So long as there were no scientific way of predicting the safety of technological structures, attempts to make devices which were new or radically different were only too likely to end in disaster. … As it turned out the old craftsmen never accepted the challenge and it is inter-esting to reflect that the effective beginnings of the serious study of structures may be said to be due to the persecution and obscurantism of the [Roman Catholic] Inquisition. In 1633, Galileo fell foul of the Church on account of his revolutionary astronomical dis-coveries … Living … virtually under house arrest, he took up the study of the strength of materials, [it] … being the least subversive subject.

(Gordon 1978: 27–28)

What Jim Gordon is alluding to here is that a new role and discipline arose around this time that was independent of and unrelated to the craftsmanship of the medieval builders. He is also quite resolute in that it could not be any other way. Within architecture, there is a roll-over effect from the Gothic master builders to the Renaissance architects, but it too suffered this chasm where new techniques gave rise to new methods, particularly in draughting and perspective views. In this, the pertinence, cohesion and comprehensiveness of an architectural response could be summarised by its quality of rigour (Burdett and Wang 1990).

History cannot be touched without changing it.

(Giedion 1944: 5)

Architecture has for far too long taught and practiced technology as an autonomous technocratic topic – often characterised by an unreflective technological determinism that results in technocratic courses, pedagogies, consultants and intellectual habits, and thereby engendering a range of problematic practices.

(Moe and Smith 2012: 2)

But not addressing these issues led Eduardo Souto de Moura to lament:

When I was … ready to go on, firemen had already defined the height, British consultants the pillars' module (three cars), the engineers the thickness of the slab. With the central core imposed by security norms, the size of the building was established by the depth of the overhanging slabs. In the end the width was 27 meters … The profile was imposed and Alberti (*utilitas, firmitas, venustas*) was definitely buried …

(Beccu and Paris 2008: 10)

Sadly, he is bemoaning the reduction of his role as architect in the Burgo Tower in Porto in 2007.

1.3 Evolution of co-ordinated project information

The intention of this work is to inform and chart what is happening in the construction sector today with regard to the impact digitalisation is having on how we make and use buildings. It offers a second-wave opportunity to re-examine the RIBA Plan of Work and to make it relevant to a large and important part of the construction industry, namely the

Figure 1.3 Office facade project

SMEs that represent the largest group of resources within building practice. Remember, in the UK:

> [t]here were an estimated 4.9 million businesses in the UK which employed 24.3 million people, and had a combined turnover of £3,300 billion. SMEs accounted for 99.9 per cent of all private sector businesses in the UK, 59.3 per cent of private sector employment and 48.1 per cent of private sector turnover
>
> (Federation of Small Businesses 2014)

and in the USA:

> The sheer number of construction firms (710,000 in 2002 – USA) and their size – only 2 percent had 100 or more workers, while 80 percent had 10 or fewer workers (CPWR, 2007) – make it difficult to deploy new technologies, best practices, or other innovations effectively across a critical mass of owners, contractors, and subcontractors.
>
> (Kennedy *et al.* 2009)

This middle ground needs a voice and guidance in how to tackle these new requirements, implement effective strategies and adopt robust methods to improve baselines, productivity and performance. To their great credit their size, contrary to the above quotes, means that they are able to adapt and respond relatively quickly to the changing sands of time. But if they are to do this appropriately, good guidance and sensible arguments need to be presented so that they willingly adopt the mantle now bestowed.

Addressing these issues is something the next generation of construction professionals will spend much time and effort amending. Nearly 25 per cent of world greenhouse gases come from electricity and heating, of which 40 per cent arises from residential buildings and 25 per cent from commercial buildings. This is about a third of the carbon dioxide emissions globally (World Resources Institute 2005). The existing building fabric needs to be retrofitted to abate these numbers.

With regard to urban growth, 10 per cent of the population lived in cities in 1900, rising to 50 per cent in 2007, while it is expected to reach 75 per cent in 2050 (Burdett and Sudjic 2008). Note that these habitats have not been built yet, meaning there is much to do. London in 1939 was the second biggest city in the world (after New York). It is just about to reach that size again (8.615 million), but it does not even make the top 20 today (Prynn 2015).

Exploitation of technology, coupled with innovation, will occupy much of the solutions tailored and matched to how we respond to the current crisis. The scale of this suggests that each iteration of innovation can deliver a technological jolt as powerful as all previous rounds combined. Economists divide growth into two types: extensive and intensive. Extensive growth is a process of adding more. This can be labour, capital or resources, which is ultimately subjected to diminishing returns. Intensive growth seeks to discover better ways to use those finite precious resources. It looks to improve technologies with a technique called 'growth accounting'. We are now entering this zone.

1.4 The changing nature of information management

BIM is basically a process where data is added. Essentially, it is form-giving, or modelling, but integral to the process is the adding of properties to this form. This allows perfect co-ordination and communicates with everything and everyone involved (Deutsch 2011). It can start with a spreadsheet merely informing a financial model or ratifying the amount of space or size required. It can end with automated sensory feedback on the facility's use through or over time. In between, it will build, grow, analyse, scrutinise and simulate all facets of the project across the whole supply chain and for all stakeholders. It will be a parallel virtual world mimicking the actual facility.

How that data is managed then becomes paramount, and this is achieved through something called metadata. Metadata, a subset of data, is latent, meaning it is there for those who need it and transparent to those who do not. Critically, filters are needed to present and view relevant data to relevant bodies, be they stakeholders, third parties or non-disclosed. The last mentioned is significant as there is increasingly more coding and machine reading that goes on without us knowing per se.

This can include apps (applications) or bots (robots) that process the information, find solutions, and approve or flag issues that would otherwise take a much longer time (Obonyo 2010). Examples of this are digital building code approval, or search engines that automatically find building products that meet an element's increasingly refined properties, through various levels of detail (LOD).

BIM is not only an authoring tool for architects and engineers, but also an analysis tool for all stakeholders in the building programme procurement process. Analysis tools, such as those that check the code of building regulations, and environmental simulations that can report on heating loads, daylighting and carbon use, will influence the adoption of intelligent modelling faster and further than previously thought.

The benefits of BIM for clients should not be underestimated either, and some clients are already reaping them where project certainty is to the fore: 'BIM, which began primarily as a design tool then evolved to a must-have for leading contractors, is now rapidly gaining traction with owners around the world' (Jones and Bernstein 2014: 4). However, the professional language that architects and engineers espouse is a latent force that can run counter to fostering collaboration. An emerging professional, the architectural technologist, can bridge that divide and adopt the adjunct role of manager in the integrated project delivery.

One of the aims of this book is to accelerate and propel an integral part of the construction industry forward by giving an insight into what is required and how to navigate a safe passage through this new way of working using building information modelling. In a nutshell, it offers a method of collaboration in a terribly fragmented sector.

It paints the broader picture of how procurement is only a part of a building's life, bringing initial strategies and sustainable mindsets to the fore. It extends the process into the everyday use of the building, post-occupancy checklists and environmental performance, giving society a better product, restoring certainty, while replacing rhetoric with reality.

The use of BIM is undoubtedly increasing, especially now, due to governmental initiatives that mandate the use of digital project delivery (Tranum Mortensen 2014). BIM facilitates this move from adversarial delivery methods to collaborative practices. This process brings clients, consultants and contractors together to share risk, liability and revenue while including third parties where relevant, be they municipalities, interest groups or users. It also brings a layer of transparency to proceedings, which sadly has been lacking previously.

Finally, BIM brings methods to the fore where analysis, simulation and performance can be measured and tested before on-site execution. This process is critical to tackling concerns regarding sustainable and addressing carbon issues: climatic consequences, orientation and alternative constructions can be analysed; sequencing, cost management and life-cycle information management can be extracted to simulate 'what-if' scenarios; and expected portrayals, returns on investment and informed benefits can be shown to alleviate fears of performance in the longer operational period of the building's life.

Figure 1.4 Western Harbour Malmö project

Even within the procurement process, better methods of conveying intent, having just-in-time, relevant, filtered information and placing the construction process are needed. This is within a context where who is doing what, what is in place before and after on the site, and health and safety fears can be identified to those who need them, in the language they understand. This is achieved with formal bureaucracy removed, so that plain language descriptions and visual communication abound.

References

Barrett, N. (2011) The Rise of a Profession Within a Profession: The Development of the Architectural Technology Discipline Within the Profession of Architecture. OpenAIR@RGU, [online]. http:// openair.rgu.ac.uk [accessed 13 August 2015].

Beccu, M. and Paris, S. (2008) *Contemporary Architectonic Envelope Between Language and Construction.* Rome: Designpress.

Bernstein, H.M. and Russo, M.A. (2013) *Information Mobility SmartMarket Report (2013) – SmartMarket Reports – Market Trends.* USA: McGraw Hill Construction.

Burdett, R. and Sudjic, D. (2008) *The Endless City: The Urban Age Project.* London: Phaidon Press.

Burdett, R. and Wang, W., eds (1990) *9H – On Rigour.* Cambridge, MA: MIT.

Deutsch, R. (2011) *BIM and Integrated Design: Strategies for Architectural Practice.* Hoboken, NJ: John Wiley & Sons Inc.

Eckblad, S., Rubel, Z. and Bedrick, J. (2007) *Integrated Project Delivery: What, Why and How,* 2nd May 2007, AIA.

Emmitt, S. (2012) *Architectural Technology.* 2nd edn. UK: Wiley-Blackwell.

Federation of Small Businesses (2014) Small Business Statistics, [online]. www.fsb.org.uk/stats [accessed 31 August 2014].

Giedion, S. (1944) *Space Time and Architecture: The Growth of a New Tradition.* Cambridge, MA: Harvard University Press.

Gordon, J.E. (1978) *Structures or Why Things Don't Fall Down.* Reprinted 1980 edn. Middlesex, England: Penguin Books Ltd.

Jones, S.A. and Bernstein, H.M. (2014) *The Business Value of BIM for Owners SmartMarket Report – SmartMarket Reports – Market Trends.* USA: McGraw Hill Construction.

Kennedy, T.C., Daneshgari, P., Galloway, P.D., Jirsa, J.O., Khoshnevis, B., Peña-Mora, F., Schwegler Jr, B., Skiven, D.A., Vanegas, J.A. and Young Jr, N.W. (2009) Advancing the Competitiveness and Efficiency of the U.S. Construction Industry [Homepage of The National Academies Press], [online]. www.nap.edu/catalog.php?record_id=12717 [accessed 4 March 2014].

Kowalski, J. (2012) Design as a Non-sequential Process [Homepage of Autodesk], [online]. au.autodesk.com/?nd=class&session_id=10961 [accessed 17 February 2013].

Le Corbusier, C.-E. J. (1921) *Towards a New Architecture.* Dover Publications, 1985.

Mayne, T. (2006) *Change or Perish Report on Integrated Practice.* USA: AIA.

Mies van der Rohe, L. (2012 [17 April 1950) Architecture and Technology (speech at the Blackstone Hotel in celebration of the addition of the Institute of Design to Illinois Institute of Technology). [Homepage of The Mies Society, MIT], [online]. www.miessociety.org/speeches/id-merger-speech/ [accessed 24 April 2013].

Moe, K. and Smith, R.E., eds (2012) *Building Systems: Design, Technology and Society.* New York: Routledge.

Obonyo, E. (2010) 'Towards Agent-Augmented Ontologies for Educational VDC Applications', *Journal of Information Technology in Construction,* 15, p. 318.

Prynn, J. (2015) 'London Population Boom: Number Living in the Capital Set to Hit All-time High Within Weeks', London *Evening Standard,* 6 January, [online]. www.standard.co.uk/news/london/london-population-boom-number-living-in-the-capital-set-to-hit-alltime-high-within-weeks-9957804.html [accessed 6 January 2015].

Tranum Mortensen, O. (2014) *Implementation of Building Information Modelling and the Impact on Design Culture*. Bachelor Thesis. Copenhagen: KEA.

Tse, T.K., Wong, K.A. and Wong, K.F. (2005) 'Utilisation of Building Information Models in ND Modelling: A Study of Data Interfacing and Adoption Barriers', *Itcon*, 10, Special Issue 'From 3D to nD Modelling', pp. 85–110, [online]. www.itcon.org/2005/8 [accessed 27 November 2008].

Underwood, J. and Isikdag, U. (2011) 'Emerging Technologies for BIM 2.0', *Construction Innovation: Information, Process, Management*, 11(3), pp. 252–258.

World Resources Institute (2005) World Greenhouse Gas Emissions. World Resource Institute.

2 BIM perspectives

Figure 2.1 MiniCO$_2$ houses, Nyborg, Denmark from east

2.1 Globalisation and the built environment

'Housing is a major contributor to CO$_2$ emissions in Europe and America today. The construction of new homes offers an opportunity to begin to address this issue' (Williams 2012: i). So begins Jo Williams' book *Zero Carbon Homes: A Road Map*. With buildings being responsible for 40 per cent of energy consumption and 36 per cent of CO$_2$ in Europe, and with the average EU resident producing 938 kg of CO$_2$ against a world average of just 308, something ought to be done.

Residents play a very significant, but sometimes erratic, role in terms of their overall CO$_2$ emissions. Depending on their behaviour, they can affect the CO$_2$ balance by up to a factor of +/−10 per cent in relation to energy consumption relative to CO$_2$. Therefore, there are potentially huge savings to be made in residents' consumption of heating, hot water and especially electricity by letting the house's architectural design and technological decor help residents to change bad habits and even rediscover old traditions. Last, but not least, capping a quota for maximum monthly CO_2 consumption leads to interesting lifestyle changes.

Innovative businesses tend to be more productive and typically can grow twice as fast as those that stagnate. Innovative economies generally are more competitive; by definition, they usually respond better to change and consequently see higher returns on their investment.

The innovative commissioning of some prototype $MiniCO_2$ houses in Denmark has demonstrated that much can be learned and extracted from their performance. How they stack up against other metrics and how they lend themselves to a new typology that can be rolled out to catalogue house builders, leaving much to be noted and examined.

The nearly Zero-Energy Building (nZEB) requirement that will apply across Europe to public buildings by 2018 and all buildings by 2020 (European Commission 2012). To achieve these standards requires not only energy-efficient building fabrics and energy production, but here the demands still cannot be met without the inclusion and assessment of carbon (McGuinness 2014). Carbon has a twofold role: the first is our responsibility to make informed decisions of the embodied carbon of the materials we deploy in our buildings, the second is to sequestrate free carbon and hold it in the fabric of our building stock (Williams 2012).

One of the biggest issues facing society today is climate change (Solomon *et al.* 2007). Globally, 2014 was the hottest year on record, and the top ten hottest years have all happened since 1997, 'undermining claims by climate-change contrarians' (Gillis 2015). The three warmest years were 2014, 2010 and 2005. 'February 1985 was the last time global temperatures fell below the 20th century average for a given month, meaning that no one younger than 30 has ever lived a below average month' (Gillis 2015).

In 1824 Joseph Fourier used a corked vase to discover that there is in fact an atmosphere surrounding the earth, which traps heat and makes the earth habitable. This discovery coincided with the start of the Industrial Revolution using fossil fuel. In 1896, Knut Johan Ångström and Svante Arrhenius discovered that burning oil and coal could affect global temperatures, but not enough to warrant concern at the time.

Moving on to the 1960s, Charles Keeling then mapped a direct correlation between the amount of carbon particles in the air and the burning of fossil fuels. Moreover, quantities of carbon in the air were increasing. Keeling's evidence was proof positive that carbon was polluting the atmosphere. In 2010, NASA declared the preceding decade to be the warmest on record. This warming of the atmosphere has somehow allowed air movement along patterns not previously noted. It means tropical air can bear more moisture and Siberian airflows move colder air over wider areas. Where the two meet heralds interesting results beyond the scope of this book. As a result, today we live in a world of extreme and erratic weather, and it is costing us heavily in footing the bill for natural disasters. The easiest reform would be to remove the US$550 billion subsidies globally for producing fossil fuels (The Economist 2015).

In 2013, insurance losses from flooding in central Europe were estimated at US$3.9 billion. The Dunally bushfires in Australia cost US$87 million. Hurricane Sandy on the East coast of America cost US$65 billion, and the Midwest Plains Drought cost US$35 billion (Climate Reality Project 2013). But oil, coal and gas companies are immune from penalties and do not directly contribute or pay for the consequences of natural disasters. Six of the top ten companies on *Fortune* Magazine's Global 500 list in 2012 were oil and gas companies, making a combined US$123.2 billion profit.

The carbon credit price, ironically, is politically suppressed (Prentice 2012) due to lobbying and increasing energy prices. There is a conflict of interest from within the oil, coal and gas industries, fuelled by the political lobbyist system. It is time to let the free market put a price on carbon, to stop fossil fuel subsidies and to cap emissions (Climate Reality Project 2013). Of the 56.6 per cent of fossil fuels being burnt, China and India's share grew from 13 per cent in 1990 to 26 per cent in 2007 (Seneviratne *et al.* 2012), and this is increasing. Sharing technologies between developed and developing countries is crucial to managing climate change effectively, efficiently and equitably. There is a preference for calling this transfer of low-carbon technologies, the transfer of 'climate technologies' (Ockwell and Mallett 2012).

Ockwell and Mallett goes on to state that there are four issues which characterise low carbon technologies:

- they are quite unique from conventional technology transfers;
- innovation capacities support the uptake of new technologies;
- the socio-technical nature of technology transitions is fundamental to how low-carbon developments can be catalysed; and
- socio-technical transitions are context specific and rely on physical, cultural and economic roles for successful adoption.

With regard to 'zero energy' and 'carbon neutral', the terminology needs both definition and clarification in plain terms, as 'embodied carbon' and 'carbon sequestration' bring a new dimension to the debate. In general, three categories can be defined: the systems addressed, the resources involved and the targets themselves. Systems can be the environment lived in, such as a building, a community, a city or even a region. Resources can be climate, carbon, CO_2, emissions, energy, fossil fuels or renewable energy. Targets can be neutral, 100 per cent, zero, free or self-sufficient.

Essentially, in construction there are two types of carbon emissions, operational and embodied. Operational carbon emission refers to carbon dioxide emitted during the life of a building. This typically includes the emissions from heating:

> Embodied carbon refers to carbon dioxide emitted during the manufacture, transport and construction of building materials, together with end of life emissions. So for example, if you are specifying concrete on a project then carbon will have been emitted making that concrete. Their emissions occur during extraction of the raw materials (the cradle), processing in a factory (factory gate), transporting the concrete to a construction site (site). This we refer to as the 'embodied carbon'.
>
> (Lockie 2014)

Sequestration, on the other hand, means capturing (or a process of removing) atmospheric carbon dioxide (CO_2) and storing it safely so that global warming is deferred. Where sequestration becomes interesting in construction terms is when building materials are derived from bio-renewables, typically timber. The carbon in these materials has been removed from the atmosphere through photosynthesis and is stored in them until it decomposes or is decommissioned.

The Climate Change Act (UK) wants to reduce greenhouse gas emissions by 34 per cent by 2020 and 80 per cent by 2050. The targets for reductions in emissions by 2022 are 12Mt CO_2/pa from homes and communities and 41Mt CO_2/pa from work places and jobs. But currently sequestration is not considered mainstream as a means of achieving those targets.

> However, the amount of CO_2 sequestered within our buildings is significant. Based on extrapolations of current trends in timber frame construction the total annual carbon

Figure 2.2 MiniCO$_2$ houses, Nyborg, Denmark from west

> sequestered within Harvested Wood Products (HWP) in construction is expected to
> increase from about 8Mt CO$_2$ in 2005 to 10Mt CO$_2$ in 2020 and 14Mt CO$_2$ in 2050.
>
> (Wilton *et al.* 2013: 12)

The accounting procedure used (and agreed at COP17) is a version of the production approach, where the benefits remain in the country where the wood is grown. BIM allows us to measure the metrics, analyse the performance and simulate alternatives before they are executed.

2.2 Information technologies, incremental and step change

Besides the technologist, and also in tandem, BIM has a need to re-empower architects and managers so that they can regain the overall command in the design process if it is to be so. We can examine the opportunities afforded by large international practices, like Frank Gehry, in re-empowering their overall commanding role, but equally so these are permeating all the way down to regular practices, and will feature (if not already) in 'retro-fit' refurbishments and 'the bread and butter efforts' of younger firms in architectural competitions. From this, it can be seen that there is a need to engage architects in the decision-making processes. Equally, there is a need for architects to empower themselves and this rides very close to leveraging them into the new methods (BIM) by showing and encouraging their adoption.

Furthermore, with regard to project deliveries, the model is and will grow to help both site architects and project managers. Augmented reality (AR) together with Geographic Information Systems (GIS) is transforming this sector. Free model viewers are also influencing their impact, making them indispensable. Levels of detail are making the collaborative process better and the communication value of visual material cannot be underestimated, which accounts for much more than previously anticipated by this author.

This will address the role the model can acquire to aid the initial design decisions, all the way through to the maintaining of the project on time and to budget. Essential to these processes is the way the model can be interrogated, analysed and simulated. This is where the model begins to take on its own responsibility in impacting design work and the processes or strategies adopted for its overall effect.

Making a building by and large is a process; it goes without saying that you cannot buy a house off a shelf. No two buildings share the same site and hence there are basic differences such as location and orientation, just to start the process. Granted, there are typologies and catalogues, but essentially every building is a new beginning. Not only are the physical buildings different, but the planning and the procurement involve differing parties, and the occupancy and the use reflect differing people and cultures.

This process is a collaborative piece of work between many stakeholders over time, expressed through phases. These phases, or work stages, are designed to establish tangible milestones through the process, to aid the exchange of predefined packages of work in return for remuneration. Increasingly, these packages are being addressed as soft landings (BSRIA 2014), after each phase, in a bid to better inform the client and to ease the process, so that everyone on-board is up to date and best informed.

This plan of the process has an imprimatur to remove the pretenders and improve transparency. By pretenders is meant the deadwood and the waste that has harangued construction, detracting from productivity and increasing the unwanted excesses. For too long construction has underperformed while being extremely wasteful of precious finite resources. In fact, poor work and double work were often encouraged as this meant double payment for the defective work to be righted. There was little incentive to get it right; it was more important to minimise your risk and maximise your profit. The business was fragmented and no one sought to rectify the situation, until Michael Latham and John Egan began voicing alternative methods and practices to bring about change (Latham 1994; Egan 1998).

2.3 International and national BIM conversations

But it was not until a method of co-ordinating the project in a centralised place allowed a single point of entry, acting as an all-consuming container or moreover an umbrella organisation holding everything together entered the stage. This is commonly known as Building Information Modelling (BIM) and it has encouraged the adoption of better working methods more than any other device in recent times (Harty and Laing 2010b).

The project-based collaborative process introduces a relatively new term, 'managed collaboration', which replaces what was previously controlled by protocols and the disciplinary silos of professional bodies who addressed their own input and interfaced with other professionals in a prepared, co-ordinated and orderly fashion. Adopting managed collaboration is timely with the emergence of BIM and Integrated Project Delivery (IPD), as well as lean construction.

The new Plan of Work from the RIBA (Sinclair 2013) harnesses this new situation and offers improved procurement and better service, innovation and design. As Darwin said, 'those who learned to collaborate and improvise – most effectively – prevailed'. Collaboration is a collective intellectual function that can be a force multiplier in an effort to reach an intended objective. In a general sense, collaboration represents a device for leveraging resources

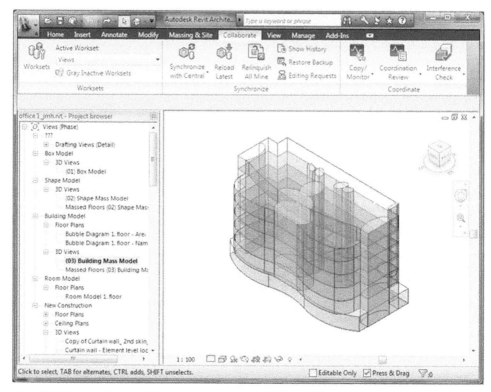

Figure 2.3 Western Harbour, Malmö project – concept block model

(Pressman 2014). In the same text, Scott Simpson goes as far as to say collaboration is an atti-tude more than a process.

Coupled with BIM being a mindset (Harty and Laing 2011), IPD comes across largely as being a method that harnesses this protocol (MacLeamy 2010), while lean construction blos-soms out of the processes (Barrett 2008) enabled in the project. But if collaborative skills and processes tend to transcend technology and tools, then the software does not of itself cultivate meaningful engagement. The force multiplier effect is achieved by the synergy that espouses from the common effort, adding richness and depth to the project. This is all in contrast to the traditional design-bid-build (DBB) method, where the architect and the contractor were natural adversaries (Pressman 2014).

On this adversarial issue, Patrick MacLeamy went as far as to say that:

> in fact today's architects spend about 75% of their time on non-design tasks, practicing what I call defensive architecture. As a result design suffers from lack of attention. Not enough time is put into thoroughly vetting the design to be sure it absolutely suits the client's purposes.
>
> (MacLeamy 2010)

This narrative comes from his famous curves talk which compared time over effort through the construction process.

Procurement now encompasses the complete supply chain and its management. Previously, a design team was assembled for the commission or project at hand and was disbanded after

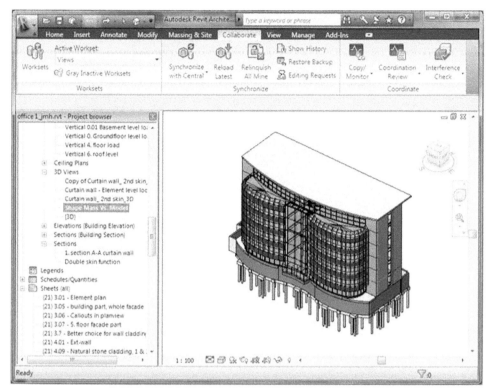

Figure 2.4 Western Harbour, Malmö project – finished model

handover. Another way of putting it is to say that upon handover risk abated drastically, barring snagging. Increasingly, design decisions are reaching beyond this ringed fence, demanding that sustainability and life-cycle assessments be taken into account. Having a 'Strategic Definition' stage together with a 'Preparation and Briefing' stage (Stages 0 and 1) before 'Concept Design' (Stage 2) allows for a considered assessment of the project to ensure that it is environmentally sustainable, which can then be fleshed out before the project itself gets under way (Sinclair 2013).

Having a new 'In Use' stage (Stage 7) begins to address the facility's management, looking at maintenance protocols and operational issues for the lifetime of the building (Sinclair 2013). This larger reach and fuller commitment is intended to make the design period more meaningful. Remaining in the loop for the design team, both before and after, is a commitment to serving the client and society better. Having a bigger picture ultimately will serve better-informed design decisions and improved certainty in design matters.

2.4 Interoperability myths and challenges

The title Mr (Ms/Mrs) rather than Dr (Doctor) is a badge of honour for surgeons, marking their delivery from subservience in earlier times. It documents the sequence of how the profession emerged from barber surgeons in the nineteenth century to the all-conquering plastic surgeons of today. So too can the branding of technologists be said to be on the cusp of shedding its image of the misconstrued subordinate role to architects. This will allow

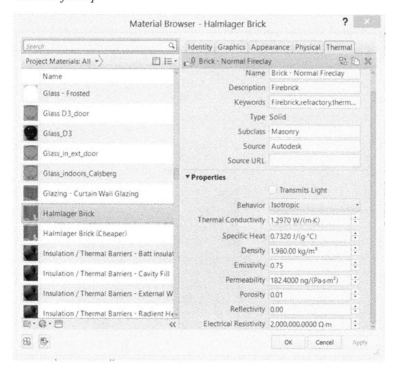

Figure 2.5 Autodesk Revit material browser with physical and thermal properties

technologists to claim their rightful place shoulder-to-shoulder with other design professionals within construction.

In the nineteenth century surgeries were placed as far away as possible from the rest of the hospital, to stifle the screams of the victims from other patients, and usually beside the mortuary, so that in the event of an unsuccessful procedure the surgeon could continue the work next door (to see what went wrong and/or improve technique) (Hollingham 2009).

The advent of anaesthetics changed all of this drastically: it pacified the patient while allowing the surgeon to focus on the job in hand. This paved the way for proper investigative methods and opened the procedures up for better techniques. For anaesthetics read 'BIM', and the potential for the technologist widens!

By branding the role of the technologist, through technology, a similar path can be taken to establish the title and build a robust and well-founded plinth upon which to promote technologists' undoubted skills. Such an action would see their meritocracy improve and consolidate their rise from technician, while also opening new specialisms. Moreover, it would profile their services and present their wares on a surer footing.

Being able to figure and document constructions makes design better informed. Analysing and simulating models makes models useful. Building information into models is the rightful domain of the technologist. BIM is the platform to make the technologist not only indispensable but critical. Modelling has the potential to reduce, even remove, prototyping from the site, consigning costly mistakes to the past while installing certainty and predictability to the much maligned sector.

'BIM is not a particular product, but rather a description of the process and intent of the deliverables used to describe, construct and even maintain a facility' (Aubin 2013). While it appears to be the implementation of a technology (Eastman 2012), it is in fact the psychology of how we can work together, collaborate with each other and trust each other's work. It is about how we share data, use data and present data in forms that are meaningful to the intended recipients. Of interest here is how this process of sharing and filtering can be accomplished and who can deliver what delivery entails (Harty and Laing 2010b). There comes a paradigm shift where levels of detail, augmented reality and Geographic Information Systems (GIS) embellish the model, making it robust and transparent but requiring solid management (Smith and Tardif 2009).

While one of the top three causes of uncertainty is unforeseen site conditions, architects and contractors point to owner-driven changes, while contractors blame design errors and omissions (Jones and Bernstein 2014). Finding a discipline competent and knowledgeable to understand digital modelling is not immediately easy; architects would like to regain their status at the top of the design team hierarchy (Gehry 2008), engineers having seen BIM's potential would like to hijack the enterprise (Throssell 2010), contractors gain better control on time and resources with its implementation (Cardenas 2008) and lastly owners and clients are beginning to see that certainty accrues from having a digital building model in place (Deutsch 2010). But while each have their well-intentioned reasons, none have grasped the gauntlet and taken ownership across the board (Harty and Laing 2011).

There is huge potential here for a discipline to brand this facet of technology (Harty and Laing 2010b). The timing of this is also well adjusted since (with regard to BIM) we are past the 'innovator' stage, which saw the emergence of digitalisation, and the 'early adopter' stage, which heralded the great strides in reforming the construction industry. We are now entering the 'early majority' phase, which will see a generally beneficial period for users and be followed by the 'late majority phase', leaving finally the 'laggards' who are in principal driven only by an overriding fear of debt (Rogers 2003).

With the title of technologist, it is befitting that the scope or range of that discipline broadens to encompass this digital phenomenon. Furthermore, the rigours of technologists' work regularly take them into close proximity with the other professions (KEA 2013). From a position of being a generalist, the time is right to exploit this situation and drive bull-headed for the finish line. All it requires is a mindset and this is well under way to being cemented through many initiatives both from within and without the profession (Harty and Laing 2011).

Regarding education, there was a problem whether to teach engineers the classics, science or merely provide training, resulting in them finally turning to the applied sciences. Enter now the architectural technologist, where these actions became even more complex, but who are traversing a similar path (AAU 2011). The comparison with surgeons relates back to the battlefield where they were the barber surgeons cutting limbs off with little scientific or technical knowledge bereft of the application of systematic research (Barrett 2011).

In the 1870s patronage was notionally abolished from the British civil service in favour of competitive entry. This was due largely to compulsory education being made available to the masses. Thus, 'merit became the arbiter, attainment the standard for entry and advancement in a splendid profession' (Young 2008: 31). In turn, 'it would be natural to expect that so important a profession would attract into its ranks the ablest and the most ambitious of the youth of the country' (Northcote and Trevelyan 1853). As a result of this, Young goes on to tell us that:

> today, we frankly recognise that democracy can be no more than aspiration and have rules; not so much by the people but the cleverest people; not an aristocracy of birth, not a plutocracy of wealth; but a true meritocracy of talent.

This emergence of the word 'talent' should not go unnoticed. Furthermore, the idea that merit should be earned should also be noted. Both of these things make knowledge and technology valued entities. In parallel fields today (IT, for example), much work has been done in Technology Acceptance Modelling (TAM) and Innovation Diffusion Theory (IDT). Essentially, they are addressing people's perception and adoption of technologies through their Perceived Usefulness (PU) and their Perceived Ease-Of-Use (PEOU) (Davis 1989). The root of the problem stems from a reluctance to embrace new technologies or to be open and accepting of foreign entities.

This behavioural trait can be a barrier to adopting new technologies for a variety of reasons, including attitude, intention and reliance. Overcoming these barriers to be seen as positive and beneficial to the user. Additionally, in marketing there is a ploy known as product placement, a method that new technologies have adopted to encourage acceptance. This is known as diffusion and is the rate by which a new idea or a new product is accepted by the market (Bass 1986).

Technologists need to adopt these methods and be seen for their worth within the construction sector. Being able to figure out and document constructions makes design better informed. Being able to analyse a construction brings certainty to the market. Being able to justify, through simulation, choices taken, puts the technologist on to the critical path.

The architectural technologist's cause is not to present the profession as a technological solution in itself, but to promote the benefits of technology. This means endorsing the return of investment, authorising case studies and proven methodologies, and diffusing the message from one stakeholder to the next, which feeds the end-users' tangible benefits for their adoption.

Ultimately, this means being able to quantify the needs, demands and requirements to enable works, rather than posing as the mere technologist equipped with the proper technologies to deliver the same. This difference is important, because if you cannot measure it, you cannot deal with it. Without empirical data there is no traceability, no accountability and, by extension, no liability to make better-informed design.

BIM is constantly evolving and, quite naturally, not everyone adopts the differing systems and technologies at a common rate. New and recent adopters will have to go through a managed process of change in their internal organisations as well as the external processes. They will also have to consider how they interface with the supply chains, clients and consultants.

Continuing to use Level 0 (maturity plan) with a lack of co-ordination increases costs by 25 per cent through waste and rework. In Levels 1 and 2, using 2D and 3D gives better probability of removing errors and reduces waste by up to 50 per cent. Under Level 3, it is then possible to reduce risk throughout the process and to increase the profit by 2 per cent through a collaborative process.

While the majority of BIM users in the UK are still working in a Level 1 process, more experienced users are seeing significant benefits by moving up to Level 2 (Bew and Underwood

Figure 2.6 Autodesk Revit model of Villa di Maser

2010). This shows that it is important to improve competences and try to reach Level 2 before the majority does, in order to maintain a market advantage. 'It is clear that organisations adopting BIM now will be those most likely to capitalise on this advantage as the market improves' (Calvert 2013).

References

AAU (2011) Architectural Design, MSc in Engineering with Specialisation in Architecture – Education – Studyguide – Aalborg University (AAU), [online]. http://studyguide.aau.dk/programmes/program/architectural-design-msc-in-engineering-with-specialisation-in-architecture(37028) [accessed 13 January 2011].

Aubin, P. (2013) How to Approach Migrating from AutoCAD to Revit [Homepage of Lynda], 8 March, [online]. http://blog.lynda.com/2013/03/08/how-to-approach-migrating-from-autocad-to-revit/?goback=%2Egde_2669081_member_221030676 [accessed 10 March 2013].

Barrett, N. (2011) The Rise of a Profession within a Profession: The Development of the Architectural Technology Discipline Within the Profession of Architecture. OpenAIR@RGU, [online]. http://openair.rgu.ac.uk [accessed 13 August 2015].

Barrett, P. (2008) *Revaluing Construction*. 1st edn. UK: Blackwell Publishing Ltd.

Bass, F.M. (1986) 'The Adoption of a Marketing Model: Comments and Observations', in: V. Mahajan and Y. Wind, eds, *Innovation Diffusion Models of New Product Acceptance*. Cambridge, MA: Ballinger.

Bew, M. and Underwood, J. (2010) 'Delivering BIM to the UK Market', in: J. Underwood and U. Isikdag, eds, *Handbook of Research on Building Information Modeling and Construction Informatics: Concepts and Technologies*. USA: Information Science Reference (an imprint of IGI Global), pp. 30–64.

BSRIA (2014) The Soft Landings – Core Principles [Homepage of BSRIA], March, [online]. www.bsria.co.uk [accessed 4 March 2014].

Calvert, N. (2013) Why We Care about BIM… [Homepage of Directions Magazine], 11 December, [online]. www.directionsmag.com/entry/why-we-care-about-bim/368436 [accessed 28 March 2015].

Cardenas, M. (2008) New ConsensusDOCS Contract First to Address BIM | AGC – The Associated General Contractors of America [Homepage of AGC], 1 July, [online]. www.agc.org/cs/news_media/press_room/press_release?pressrelease.id=192 [accessed 13 January 2015].

Climate Reality Project (2013) Home | Climate Reality, 13 March, [online]. http://climaterealityproject.org/ [accessed 13 January 2015].

Davis, F.D. (1989) 'Perceived Usefulness, Perceived Ease of Use, and User Acceptance of Information Technology', *MIS Quarterly*, 13(3), pp. 319–340.

Deutsch, R. (2010) Notes on the Synthesis of BIM: AECbytes Viewpoint #51, [online]. www.aecbytes.com/viewpoint/2010/issue_51.html [accessed 8 April 2010].

Eastman, C. (2012) Building Information Modeling 'What is BIM?' [Homepage of Georgia Tech], 3 December, [online]. http://bim.arch.gatech.edu/?id=402 [accessed 3 December 2012].

Egan, J. (1998) *Rethinking Construction: The Report of the Construction Task Force*. UK: Department of the Environment, Trans Construction Task Force.

European Commission (2012) EU Energy Performance of Buildings' Directive. Commission Directive 2012/27/EU on energy efficiency. Brussels: European Commission.

Gehry, F. (2008) Digital Project – Frank Gehry, 26 November 2008.

Gillis, J. (2015) '2014 Notches Ominous Mark as Earth's Hottest Year on Record', *International New York Times* News (41008), 17/18 January, Front page – page 3.

Harty, J. and Laing, R. (2010a) 'Removing Barriers to BIM Adoption: Clients and Code Checking to Drive Changes', in: J. Underwood and U. Isikdag, eds, *Handbook of Research on Building Information Modeling and Construction Informatics: Concepts and Technologies*. 1st edn. Hersey, NY: Information Science Reference (an imprint of IGI Global), pp. 546–560.

Harty, J. and Laing, R. (2010b) 'The Management of Sharing, Integrating, Tracking and Maintaining Data Sets, Is a New and Rather Complex Task', 2010, 14th International Conference of Information Visualisation (IV 2010), p. 610.

(2011) 'Trust and Risk in Collaborative Environments', in: J. Counsell, F. Khosrowshahi and R. Laing, eds, *Information Visualization, Visualization of Built and Rural Environments*, 13–15 July 2011, Institute of Electrical & Electronic Engineers Computer Society, pp. 558–563.

Hollingham, R. (2009) *Blood and Guts: A History of Surgery*. USA: St. Martin's Griffin.

Jones, S.A. and Bernstein, H.M. (2014) *The Business Value of BIM for Owners SmartMarket Report – SmartMarket Reports – Market Trends*. USA: McGraw Hill Construction.

KEA (2013) Bachelor of Architectural Technology and Construction Management [Homepage of Copenhagen School of Design & Technology], January, [online]. www.kea.dk/en/programmes/bachelor-degree-ba-programmes/ba-of-architectural-technology-and-construction-management/ [accessed 3 September 2013].

Latham, M. (1994) *Constructing the Team: The Final Report of the Government/Industry Review of Procurement and Contractual Arrangements in the UK Construction Industry*. London: HMSO.

Lockie, S. (2014) Embodied Carbon – a Q&A with Sean Lockie | Faithful+Gould | UK & Europe, [online]. www.fgould.com/uk-europe/articles/embodied-carbon-q-sean-lockie-director-carbon-and-/ [accessed 30 May 2014].

McGuinness, S. (2014) nZEB – The Biggest Construction Project in Human History, KEAWeek, 5–9 May 2014, Copenhagen School of Design & Technology.

MacLeamy, P. (2010) The Future of the Building Industry (3/5): The Effort Curve [Homepage of HoK], February, [online]. www.youtube.com/watch?v=9bUlBYc_Gl4 [accessed 10 September 2011].

Northcote, S.H. and Trevelyan, C.E. (1853) Report on the organisation of the civil service, together with a letter from the Rev. B. Jowett. q/JN 428 NOR. London: Her Majesty's Stationery Office.

Ockwell, D. and Mallett, A. (2012) 'Introduction', in: D. Ockwell and A. Mallett, eds, *Low-Carbon Technology Transfer: From Rhetoric to Reality*. Abingdon, UK: Routledge, pp. 3–19.

Prentice, D. (2012) *Carbon Management*. Course Material. Coventry: European Energy Centre, Centro Studi Galelleo.

Pressman, A. (2014) *Designing Relationships: The Art of Collaboration in Architecture*. New York: Routledge.

Rogers, E.M. (2003) *Diffusion of Innovations*. 5th edn. New York: Free Press.

Seneviratne, S., Nicholls, N., Easterling, D., Goodess, C., Kanae, S., Kossin, J., Luo, Y., Marengo, J., McInnes, K. and Rahimi, M. (2012) 'Changes in Climate Extremes and their Impacts on the Natural Physical Environment: An Overview of the IPCC SREX Report', EGU General Assembly Conference Abstracts 2012, p. 12566.

Sinclair, D. (2013) *Assembling a Collaborative Project Team: Practical Tools including Multidisciplinary Schedules of Services*. London: RIBA Publishing.

Smith, D.K. and Tardif, M. (2009) *Building Information Modeling: A Strategic Implementation Guide for Architects, Engineers, Constructors and Real Estate Asset Managers*. Hoboken, NJ: John Wiley & Sons Inc.

Solomon, S., Qin, D., Manning, M., Chen, Z., Marquis, M., Averyt, K., Tignor, M. and Miller, H.L. (2007) IPCC, 2007: Climate change 2007: The physical science basis. Contribution of Working Group I to the fourth assessment report of the Intergovernmental Panel on Climate Change. S.D. Solomon (ed.).

The Economist (2015) 'Energy and Technology: Let There Be Light', 17 January, [online]. www.economist.com/news/special-report/21639014-thanks-better-technology-and-improved-efficiency-energy-becoming-cleaner-and-more [accessed 30 January 2015].

Throssell, D. (2010) Minutes of interview at Royal Institute of British Architects, Friday 30 July 2010.

Williams, J. (2012) *Zero Carbon Homes: A Road Map*. Oxon, UK: Earthscan from Routledge.

Wilton, O., Robson, D., Atkinson, N. and Hill, C. (2013) *Carbon Sequestration: From Forests to Buildings*. UCL Institute for Sustainable Resources.

Young, M. (2008) *The Rise of Meritocracy*. 11th edn. New Brunswick, NJ: Transaction Publishers.

3 BIM in context

Figure 3.1 Autodesk Revit model of Tempieto Palladio

3.1 Digitisation and the third industrial revolution

The construction industry is accepted as being fragmented (Smyth and Pryke 2008), with specialists for whom it is advantageous to operate alone, bidding for individual projects and occupying a niche which can be guarded and cherished, making the whole process quite conservative and reluctant to change (Smyth and Pryke 2006). But being lightweight also makes it easier to adapt and adjust to the peculiarities of each project if the conditions are right (Sigurðsson 2009). There are also many main contractors for whom it is easier to outsource aspects of contractual work rather than carry the risk that might be associated with employing someone, and paying overheads, just to be able to offer that particular service or deliverable. This firmly bases the construction industry in pre-industrial times (Kristensen 2011).

The pre-industrial times charted the dominance of the master builder, supported by crafts-men and apprentices, who operated in a relatively closed framework of building methods and practices. Masons were masons, carpenters, carpenters and so on, each knowing what was required through a learned process of apprenticeships with expectations that were essentially ring-fenced. This clearly served the society of its time, as witnessed by the calibre of buildings produced then.

With the advent of steel and reinforced concrete came innovative developments that fell outside the confines of the previous regime (Lemoine 2006). This led to designers having to ground their designs, requiring larger teams of experts making qualified decisions. It heralded the role of the engineer in building procurement. These included structural requirements, ser-vice installations and the emergence of performance-related tasks. In 1865, the Massachusetts Institute of Technology (MIT) emphasised teaching students to use machine tools rather than finished products (Mack 2005).

Traditionally, the drawing office produced the drawings from which the building was made, together with specifications, schedules and quantities. The management, leadership and/or project co-ordinator, usually an architect, integrated these tasks and administered the contract to procure the building. The drawing process would be executed along the lines of work stages, resulting in remuneration packages for reaching preordained milestones along the process.

The work stages also reflected the movement from the conceptual sketches through to the specific details required to fashion the building, as demonstrated in the use of scales across the drawing set, starting with large scales to capture the general arrangements, right down to the small scale (even one to one) to represent the tiny details. Parallel to this, the written docu-mentation would follow similar lines, and a package would be produced for tendering which would provide the basis for delivering the building.

Christopher Alexander, in *Notes on the Synthesis of Form*, describes three scenarios of design-ing content and form (Alexander 1964). The first occurs in unselfconscious societies where the building process has remained unchanged through many generations. In this scenario the content relates directly to the form, since the person building it lives in it. This invariably means that the community has established fixed workable solutions. The situation requires no formal plans or abstracted sets of codified instructions, because methods are handed down from generation to generation without change or development. Examples of this can readily be seen in primitive tribes and even in traditional vernacular regionalism.

The second scenario happens when artisans or craftsmen emerge to do specific tasks within the community. Since it is not their house, repairs and chances for mistakes become possible. This is because there is a distinction between the doer and the user. Now that there are two or more parties there has to be a chain of command resulting in a handover of requirements. Depending on the competences of those involved, there has to be a set of codified instruc-tions, even if at a primitive level. The formal nature of this situation opens up the possibility of mistakes or errors. This is not due to any lack of quality in the work but largely because of the increase in the magnitude and complexity of the work. Alexander sees this as a semi-conscious state where the way the work is done is through an image of the content required together with an image of the form delivered. It is akin to the movement from word of mouth to the written word.

The last scenario is a formalisation of this process where the images are formalised (a for-mal image of content and a formal image of form) so that they can be better recognised and controlled. This is the fully conscious state and the building industry essentially endorses this method with formal procedures for checking and controlling the work that procures a house or whatever.

Once the act of 'drawing' was formalised, its growth possibly blossomed with the Renaissance where there was complete separation between the drawing and the project on site. But unlike the Gothic era just previous to this, the drawings began to be instances rather than types of the projects, and this abstraction saw developments in proportion and perspective being allowed to thrive. Even the term 'Gothic' is seen as derogative, with Banister Fletcher (Vasari 1991) citing Giorgio Vasari using the term as rude and barbaric back in the 1530s. This discourse continues right up to the modern movement where there is a close dialogue, and stark delimitation, between the architect and the craftsman through defined boundaries and clear expectations, culminating with nineteenth-century catalogue-type building parts.

Ironically, Adolf Loos (1930), maybe somewhat ahead of his time, had this to say:

My architecture is not conceived by drawings, but by spaces. I do not draw plans, facades or sections … For me, the ground floor, first floor, do not exist … There are only interconnected continual spaces, rooms, halls, terraces … Each space needs a different height … These spaces are connected so that ascent and descent are not only unnoticeable, but at the same time functional.

Recently, David Ross Scheer, in his book *The Death of Drawing: Architecture in the Age of Simulation*, wrote:

This long tradition of drawing in architecture, with its influence on the thinking of architects and on the very nature of architecture, is in question for the first time since the Renaissance. The divorce of design and construction, theorised by Alberti and realised in modern practice, is being overthrown by the replacement of drawing by simulation. Whereas drawing is based on a clear distinction between the two, simulation strives to eliminate any space between them. Whereas architectural drawings exist to represent construction, architectural simulations exist to anticipate building performance.

Unlike representation, which rests on a separation between a sign and reality to which it refers, simulation posits an identity between itself and reality. The nature of a simulation is that it mimics the behaviour of some real system. To the extent that it does so, it can be taken as its equivalent in the operational sense. In the limit, simulation becomes indistinguishable from reality. That it rarely actually does so does not change this fundamental relationship. One approaches a simulation with the expectation that it will produce the experience of reality to some extent. Interpretation plays no part in understanding a simulation – the experience of it is taken at face value. Note that it is experience that simulation reproduces – there is no depth behind the experience, no deeper meaning to be found. Thus, with simulation reality is identified with outward behaviour rather than with some kind of fundamental nature.

(Scheer 2014: 9)

Paolo Belardi, in his work *Why Architects Still Draw* (2014: ix) likens a drawing to an acorn, where he says: 'It is the paradox of the acorn: a project emerges from a drawing – even from a sketch, rough and inchoate – just as an oak tree emerges from an acorn.' He tells us that Giorgio Vasari would work late at night 'seeking to solve the problems of perspective' and he makes a passionate plea that this reflective process allows the concept to evolve, grow and develop. But without belittling Belardi, the virtual model now needs this self-same treatment where it is nurtured, coaxed and encouraged to be the inchoate blueprint of the resultant oak tree.

Figure 3.2 Day Care Centre project – courtyard

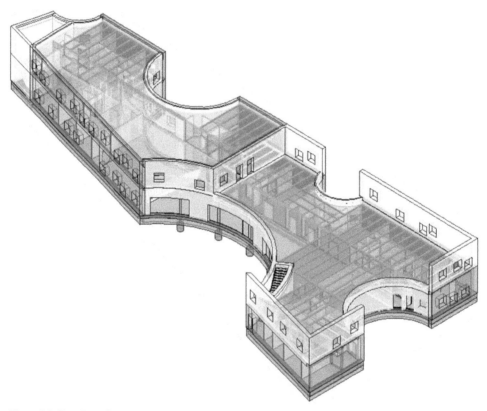

Figure 3.3 Day Care Centre project – structure

The model now too can embrace the creative process going through the first phase, preparation, where it focuses on the problem. The manipulation of the available material can then be incubated so that it is reasoned and generates feedback. The third phase, illumination, is the epiphany of the solution as Graham Wallas described it in 1921. Finally, the fourth phase, validation, makes the idea comprehensible, communicable and feasible (Wallas 1921; Belardi 2014).

One area where modelling becomes indispensable is where communication is paramount. The visual nature of a model is untouchable in its ability to show intent, flaws, benefits, or whatever. This ability to transcend an area where drawings fail, in layman's terms, whether in legal and contractual circles or on site, is disingenuous at worst. How often has a client's jaw dropped at handover, manifesting the sheer bewilderment of just how the project materialised through a mass of drawings and documents? BIM corrects this shortfall.

3.2 The Gehry studio, technology transfer and joined up process

Frank Gehry in the Digital Project Exhibition (2008) said:

> The culture of architecture in our (or in my) time works like this; you do a job; you meet a client, they hire you to do a project … Of course, they have a budget, which they tell you and a time schedule or whatever. So you finish the design and you put it out to bid, and then it comes in over budget. That (happens), I'd say, 80% of the time … Of course, the owner finds himself very confused about this, for the most part, because they don't have the extra million dollars or whatever it is, and they're on the way or they're underway, and it's very hard to stop or be sympathetic to the architect, or to the project. They feel betrayed, and this happens all the time, and it's an uncomfortable place to be but no matter how much work you do, an architect can't control the marketplace … Now you can be as careful as possible about working for budgets but I've always hated that moment, and my friends have always hated that moment and you sort of wonder is there some way out. In the middle ages the architect was a master builder, they built the cathedrals, they were respected, they had a process and it was done over centuries so no one got the blame [laughs]. In our time you have the Sydney Opera House where poor Jørn Utzon gets clobbered. It's a horrible story. It practically destroyed the man's life.

To back this up with hard examples, one needs look no further than the Walt Disney Concert Hall at 111 South Grand Avenue in downtown Los Angeles, California, and the Guggenheim Museum, Bilbao, along the Nervión River in downtown Bilbao, Spain. They are sequential projects where the first, the Disney Hall, cost $274 million, but was over five times the budget and late, while the Guggenheim steelwork bids alone came in at 18 per cent under budget. In round figures the difference is five times over budget to one-fifth under budget.

With regard to the concert hall, Gehry found himself beset with cost overruns so that the project was shelved for a long period due to lack of funding. It finally cost an estimated $274 million, which is more than five times the $50 million budget at the start of the project. Without the prestige of this client, it would probably have been shelved or watered down immeasurably.

Figure 3.4 Day Care Centre project – perspective

In this situation, Gehry has said that his position went from having the parental role at the start of the project where he was in control, to an infantile one when cost overruns threatened to scupper it. The focus moved from the architect to the contractor. The architect had lost face in the eyes of the owner, and the contractor was now seen as the saviour, if the building was to be realised. Conversely, when tendering came about for their next commission, the Guggenheim Museum in Bilbao, Gehry Technologies (GT) sent a member of staff over to Bilbao to train the bidders in the software prior to tender, which was pretty unheard of in 2004. The result was they came in under budget, seeing roughly one-fifth being knocked off the expected estimate.

How can one project with conventional tendering end up five times over budget and arguably the other, with a common model, come nearly one-fifth under budget? The upshot is that subsequently people who wish to work with Gehry must adopt his processes and pre-qualify for collaborative work. It has put Gehry firmly back in the parental role, at the helm of the ship, where he is in control. It has heralded a new dawn for Gehry where he now uses selective tendering, and bidders learn how to extract quantities. The intelligent model (BIM) has done this for him.

One of Gehry's latest projects to be realised is the Louis Vuitton Foundation for Louis Vuitton Möet Hennessy in Bois de Boulogne, in the sixteenth arrondissement of Paris, France. It was prepared largely paper-free (i.e. digitally), with a budget of $143 million (Erlanger and Gohin 2011). It consists of several icebergs rising from the ground encased in curved glass sails, held in place by an elaborate timber structure.

Once the geometry was captured by GT in Computer Aided Three-Dimensional Interactive Application (CATIA), it was Computer Numerical Controled (CNC) to suppliers and manufacturers, where through collaboration and teamwork the form was optimised and the parts produced. Upon placement, laser scanning allowed the as-built scenario to be mapped into the model again for clash detection to check if the model or the site was correct. The model remained up to date during the entire process. At handover, an as-built model can be provided to the facilities management team through a seamless operation.

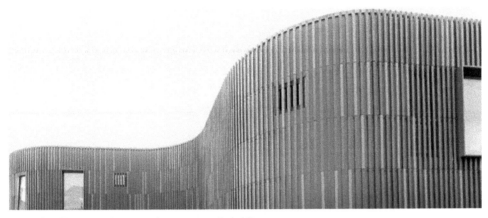

Figure 3.5 Day Care Centre project – external cladding

3.3 Heathrow T5, a tipping point for BIM in the UK?

Heathrow Terminal 5 was a Herculean project, not only for the exemplary superstructure itself, but also for the vast infrastructure required to service it. It cost £4.3 billion and even went through 46 months just to come through the longest public inquiry in British history. The Longford River had to be rerouted, there were two new tunnels (of nine), 'HexEx Bored' and 'PiccEx Bored', which were directly under the terminal for the Piccadilly underground line and its station. There was a need for reversible sidings, as well as heavy freight rail provisions close to the perimeter and various other connections to the satellite terminals. It is reported that the Accident Frequency Rate (AFR) was under 25 per cent of the national level (Ferroussat).

David Ferroussat (2007), Commercial Quality & Resource Leader at BAA, reports that large projects usually go wrong, citing:

- London Underground's two-year Jubilee Line Extension delay;
- Railtrack's missed deadline to increase slow train paths on the West Coast Main Line;
- the Millennium Dome's financial crises; and
- the British Library's failure to meet the challenges of the internet age.

To correct these imbalances, he claimed that processes, the organisation and, more importantly, behaviours should be designed to expose and manage risk, promote and motivate opportunities and address performances in all relationships. Furthermore, leaders were to recognise that change and uncertainty were the new norm and that a different outcome meant doing something, precisely that, differently.

This was achieved through the special contract (T5 Agreement), vigorous health and safety demands, high standards of quality (behavioural approach) and using milestones to apply handover pressure in the programme. The contract was a unique legal document that managed the cause and not the effect, ensured successes in a very uncertain environment and focused on managing the risk rather than circumventing litigation. There was an incentive fund to replace normal risk payments, which funded shortfalls and provided opportunities to increase profits. Finally, the project, and not the suppliers, was insured against damage to property, injury, death and professional indemnity.

Figure 3.6 Terminal 5 Heathrow Airport structure

Because of carefully defining responsibility, accountability and liability, the focus became delivery. Remuneration was based on reimbursable costs plus profit with a reward package for successful completion. This incentive plan encouraged exceptional performance with the focus on the issues of value and time. Value performance occurred primarily in the design phases and was measured by the value of the reward fund for each Delivery Team and calculated as the sum of the relevant Delivery Team Budget less the total cost of the work of that Delivery Team.

The time reward applied only during the construction stages. Here, worthwhile reward payments were available to be earned for completing critical construction milestones early or on time. If the work was done on time, a third went to the contractor, a third went back to BAA and a third went into the project-wide pot that would only be paid at the end (Smyth and Pryke 2008). There was a no-blame culture, meaning that if work had to be redone the fault was not apportioned to anybody, but the rewards would either be reduced or not awarded at all. This had the effect of applying a kind of peer pressure where it was in the interest of all parties not to fail, which created a place where the vertical silos of expertise were traded for viaducts of collaborative techniques. BAA took out a single premium insurance policy for all suppliers, providing one insurance plan for the main risk. The policy covered construction and Professional Indemnity (Potts 2009). The overall supply chain was pyramidal, with 80 key first-tier suppliers, around 100 other first-tier suppliers and thousands of second-tier and other suppliers.

While the T5 Agreement handbook sought to lay down binding guidelines for the whole supply chain during the procurement of the facility, it also went to great lengths to be readable for all involved and understandable in its holistic approach. It clearly defined and set out the expectations for everyone. It was ambitious, with two overriding standards: how to deliver and what is actually delivered. Best practices were benchmarked, levels of performances were outlined and expectations were raised across the enterprise. Three levels were identified; 'business as usual', which was rejected as a non-starter, 'best practice', which received an amber light and 'exceptional performance', which was seen as world class, receiving the green light and setting the bar.

The mission was to deliver an airport through teamwork while maintaining and delivering a strong sense of personal identity and achievement for all involved. The teams ranged from client teams through to suppliers, and from management to trainees, all identified as an integral part of the supply chain. Emphasis was placed on (pre-) planning the requirements and assigning the best resources to accomplish them. Team building and its environs were cherished in an environment designed to break down (legacy) barriers and divisions. In the longer term, this impacted social and non-work relationships, because collaboration and the building of interdisciplinary trust were emphasised.

Appropriate training was tabled as being critical. Responsibilities, and how relationships were developed, were dealt with through ingeniously defining roles and relationships as being open, questioning and non-perspective. There was a desire to match authority with responsibility, empowering people and encouraging delegation. It was essentially a framework that had not been tried before. It was also setting out the limits and parameters to which the exceptional goals could be reached. Under behaviour, colleagues were to be treated as customers, personal performance was challenged, initiative and leading by example was encouraged, and proactive positive mindsets were seen as vital. Problems were to be dealt directly, being flexible while accommodating all contributions. All of this was to be demonstrated with proper documentation and measurement.

Finally, recognition and reward were seen as great motivators in this postmodern relationship context. This was definitely the carrot rather than the stick driving the changes. In an interview with David Ferroussat, Commercial Quality & Resource Leader at BAA, the implementation of these principles typically saw subcontractors having to share sensitive information. An example would be if one subcontractor could source certain materials cheaper or better than another. The set-up meant that the subcontractor who could get the best deal supplied all, even if this meant giving a competitor (outside the contract) insight into its coveted methods and honed procedures.

Often this would lead to an impasse where there would be blatant refusals to comply ('more than my job's worth' scenarios), which could only be resolved by taking each subcontractor's bosses upstairs, quantifying the risk and potential loss, and paying out on the information, so that the 'open' dialogue could flourish. This was unheard of in the construction industry before, and for such an arrangement to flourish, a longer-term relationship is required. BAA could provide this environment with the lure of further contracts, on account of the size and scope of its organisation. This is significant while a state of transition exists.

Furthermore, when work was rejected or needed to be redone, there was a no-blame culture in place. This meant one subcontractor could not point the blame at another, but rather both had to submit proposals to rectify the work and correct it. The quicker and sooner that these things could be accomplished the better, because such extra work was paid out of the golden egg lump sum for finishing the work on time and to date. The longer subcontractors bickered, the more of the sum was eaten away. There was no incentive in reducing the bonus.

This was further enhanced with a critical path identifier, in the form of an object, a lump of rock, called a 'milestone', which would reside with the current critical deed, like a hot potato. The quicker the owner could pass the object on, the better, as there was a kind of a peer pressure culture established, encouraging the completion of tasks successfully to further the job and come closer to nirvana: practical completion.

So much for the management structures, on the technical side Terminal 5 was procured using Autodesk Architectural Desktop (ADT), predating Revit, which meant that the fully immersive milieu of sharing models and data did not happen in large amounts, but there was heavy involvement of NavisWorks to aid this aspect of the project (Lion 2004). The 3D co-ordination used NavisWorks as a process checker to view, review, detect clashes and extract

Figure 3.7 Terminal 5 Heathrow Airport concourse

information from the model. There was an estimated 10 per cent saving in design time and better co-ordination.

All of the above merely co-ordinated the 3D geometry, with all output being 2D (plans, sections and elevations) extractions. This was to radically change with the advent of BIM programmes such as Autodesk Revit where data could be added to the geometry. 2D extraction is often still a legal requirement but increasingly the model is gaining in stature.

3.4 Fast-forwarding from *Rethinking Construction* BIM analogies

Two major reports from the 1990s in the UK set the scene, which began the reforms and mapped the process to change the construction industry both nationally and with increasing influence across the globe. The first, *Constructing the Team* by Latham, a former Conservative MP, recommended a need for better standards in construction contracts at a time when there was little or no cross-platform uniformity. He called for better guidance on best practices and legislative changes towards arbitration in an attempt to change industry practices, 'to increase efficiency and to replace the bureaucratic, wasteful, adversarial atmosphere prevalent in most construction projects at the time' (Latham 1994: 45).

The report wished to delight clients (Latham's words) by promoting openness, co-operation, trust, honesty, commitment and mutual understanding among team members. Incredibly, all of these aspirations have remained on the agenda right up until and including today. Finally, he identified and determined that efficiencies, especially in savings of the order of 30 per cent, were possible over five years. In the report, he condemned existing industry practices as being ineffective, adversarial, fragmented, incapable of delivering for its customers and lacking respect for its employees. Even at this early stage, he urged reform in the industry and advocated partnering and collaboration by construction companies. He went on to say: 'Partnering includes the concepts of teamwork between supplier and client, and of total continuous improvement. It requires openness between the parties, ready acceptance of new ideas, trust and perceived mutual benefit' and 'Partnering arrangements are also beneficial between firms' without becoming 'cosy' (p. 68).

Of the recommendations in the report, the two most notable with reference to this book are:

- 'The use of co-ordinated project information should be a contractual requirement.'
- 'The role and duties of project managers requires to be more clearly defined.'

Following on from this seminal report came the Egan (former chief of Jaguar) Report, *Rethinking Construction* (Egan 1998), which identified five 'drivers' to improve construction practices:

- committed leadership;
- a focus on the customer;
- integrated processes and teams;
- a quality driven agenda; and
- commitment to people.

It identified a further four processes as needing significant improvement:

- product development;
- project implementation;
- partnering the supply chain; and
- production of components.

Finally, it called for a set of targets to be improved by certain percentages:

- capital costs were to be reduced by 10 per cent;
- construction time was to be reduced by 10 per cent;
- predictability was to be increased by 20 per cent;
- defects were to be reduced by 20 per cent;
- accidents were to be reduced by 20 per cent;
- productivity was to be increased by 10 per cent; and
- turnover and profits were to be increased by 10 per cent.

It also called for decent and safe working conditions, and an improvement in management and supervisory skills. This was coupled with long-term relationships based on clear measurement of performance and sustained improvements in quality and efficiency. That this has been so successfully undertaken with regard to health and safety shows that with the right legislation it was achievable. The best embodiment of these works and the culmination of Egan's work could be said to be encapsulated in Terminal 5 at Heathrow,

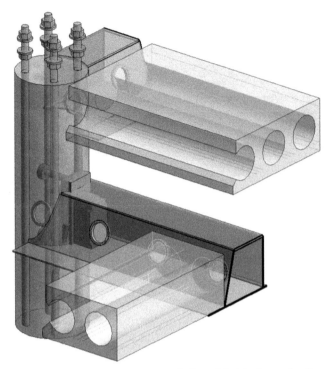

Figure 3.8 Day Care Centre project hollow slab delta beam detail

opened on 27 March 2008. Egan, chief executive of BAA, commissioned the terminal and implemented the *Terminal 5 Agreement: The Delivery Team Handbook* (Haste 2002).

The SPICE project (Structured Process Improvement for Construction Enterprises) in trying to meet the challenges of the Latham and Egan reports, looked at addressing the issues that the industry had shortcomings with a view to securing improvement. Because of the absence of guidelines, benefits could not be adopted across the board. For any work that had been done in these areas, it was difficult to assess the improvements or resources appropriately. This ultimately meant it was not possible to benchmark or measure performance across organisations.

Moreover, *BIM and Construction Management* by Hardin is a very practical oriented book (Hardin 2009). The subtitle *Proven Tools, Methods and Workflows* is, essentially, what it delivers. It presents an array of practical information aimed at the user. It has tried and tested methods of binding the 3D geometry with the fourth dimension (time) together with the fifth dimension (identified as resources), which can be assembled together in programmes such as NavisWorks. Here clashes and collisions, as well as timeline monitoring, can be mapped and resolved before becoming an on-site problem, as was the case traditionally.

Hardin discusses the parameters of management, pre-construction, construction administration, sustainability and Facilities Management (FM). Indeed, this is one of the first publications where FM is seen in practice as an intrinsic part of the building process (life-cycle assessment), which now is a critical part of the whole process. The total integration is also well documented, where he shows that he is indeed on top of where the whole process is going. As can well be imagined, the integration of these differing technologies into the traditional work phases is new and can be daunting. Hardin gives step-by-step guidance in a straightforward manner.

A complementary publication, *BIM Handbook: A Guide to Building Information Modeling for Owners, Managers, Designers, Engineers and Contractors* by Eastman *et al.* (2008) covers all aspects of BIM, from theory to practice, for all stakeholders, including a look at the future together with a selection of case studies to illustrate the work and identify good practices. It can be read from cover to cover or dipped into for relevant chapters or sections. It is a comprehensive collection of knowledge on this relatively new approach to design, construction and Facilities Management.

While strong in areas of life-cycle assessment and contractor involvement, it is weak in that it fails to identify just how the collaboration that is critical to the process is achieved and maintained. But it does correctly identify the two main drivers of client demand and productivity gains that we can see becoming mainstream in the next few years. It also sees the lack of trained personnel as the biggest barrier to adoption.

Following Eastman *et al.*, Smith and Tardif's book *Building Information Modeling: A Strategic Implementation Guide for Architects, Engineers, Constructors and Real Estate Asset Managers* (2009) provides a more strategic coverage of the implementation of BIM. Cultural changes mentioned include building trust and mitigating risk, which Eastman *et al.* do not address as comprehensively. Also, FM and life-cycle assessment feature more prominently, giving the book more purpose; it also is not afraid to name names, such as buildingSMART® and FIATTECH, to make its point.

In the seminal publication *BIG BIM little bim: The Practical Approach to Building Information Modelling* by Jernigan (2007), great emphasis is placed on the process and culture rather than the technology and software. That said, it stresses the importance of the guidance aspect associated with BIM and how the process is actually delivered. It identifies four phases for integration: initiate, design, construct and manage, with certainty as its mantra. It is firmly based in the design team issues, rather than the wider contractor and subcontractor benefits or the life-cycle issues for owners.

Life-cycle assessment and sustainability are addressed by Krygiel and Nies (2008) in *Green BIM: The Successful Sustainable Design with Building Information Modeling*. Notwithstanding its focus on sustainability, it goes deeply into the underlying issues of BIM, IPD and methodologies before delving into sustainable forms and systems. These are addressed with orientation, massing and daylighting for the questions of form, with water harvesting, energy modelling, renewal energy and sustainable materials for the systematic side of things.

In conclusion, Krygiel and Nies call for:

- more interoperability between software packages, a lack of which is seen as a great drawback;
- more input from the designer in the modelling;
- a carbon counting method in the form of a dashboard on the fly through the design phase;
- more immediate responses to calculations of rainwater to roof areas, window to wall ratios; and
- better interactivity with climate data.

All of which is eminently available now and possible because BIM is essentially a database. Taking perhaps a more managerial approach, Elvin places the emphasis on the benefits and cultural change required to effect collaboration, in *Integrated Practice in Architecture, Mastering Design-Build, Fast-Track, and Building Information Modeling* (Elvin 2007). The book remains solely in the architectural domain and does not venture into the whole supply chain. Its ethos can be seen in the AIA's stance where architects see integration as belonging in their function as the natural team leader.

This is also taken up with *Design Management for Architects* by Emmitt, who looks at the broader picture and makes the case for architects retaking the role as the managers (Emmitt 2007). Coupled with this is *Architectural Management*, which documents the added value from good management, both in design, communication, integrity, practice and education (Emmitt *et al.* 2009). Critically, Emmitt laments the level of management undertaken at schools of architecture and comments on the knock-on effects such will have.

Predating digitalisation, in its scope presented here, Winch provided some useful insights into the politics (small 'p') of collaboration and large project consortia in his book, *Managing Construction Projects: An Information Processing Approach* (2002). There is a good account of the roles and how they dovetail with each other. Vignettes are drawn of how real projects developed through the roller coaster modes of management back then, in a matter-of-fact, documentary style, highlighting, for want of a better word, the need for the change that the industry is now experiencing.

Finally, collaborative relationships, both in contractual frameworks, conceptual frameworks and through networking, are well covered in *Collaborative Relationships in Construction: Developing Frameworks & Networks* by Smyth and Pryke (2008). Smyth starts by saying value is added to

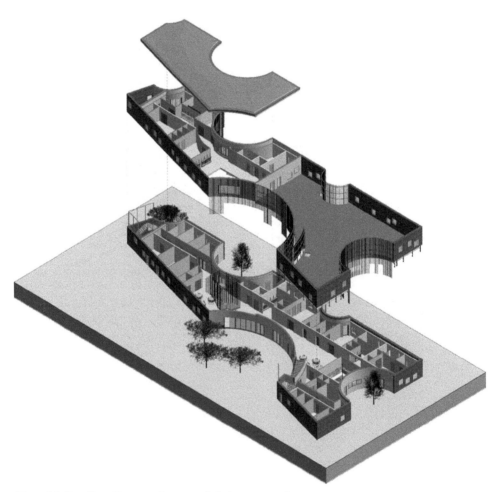

Figure 3.9 Day Care Centre project – exploded axonometric

projects through people, and this drives the remainder of the work. While acknowledging traditional methods, as well as information processing and functional management, the basis of the book explores relationship approaches. This opens a socio-psychological perspective, which drew me beyond my first boundaries for this book. It takes partnering further than before laying the groundwork for longer-term relationships. It takes the roles further than being project delimited, meaning that people are beginning to look beyond single-project handovers, wishing to build relationships where there is trust, and where they use each other for what they are best suited, while sharing and reaping the many benefits of such illicit affairs (Smyth and Pryke 2008).

The book refers back to Terminal 5 and much of the work draws from what happened there. It explored new ways of collaboration, with emphasis on the trusting element that is necessary in any relationship. Indeed, it was like a formalised speed-dating session where both, or all parties, needed to establish their credentials, together with a method of finding out how compatible they were with each other and also finding ways of linking like-with-like so that there was not an imbalance in the relationships formed.

In terms of papers, reports and other research, numerous texts come to mind, which have been grouped under phases. The first phase happened when BIM first formalised itself, broadly speaking when Jerry Laiserin first coined the term in his letters' column, 'Comparing Pommes and Naranjas'. He notes about having lengthy discussions in getting the two top-weights, Autodesk and Bentley, to agree on the common term, BIM, as a generic term similar to CAD. The same series notes the development and strategy of Autodesk viewing Revit as the obvious platform for the building industry, saying: 'Revit is Autodesk's strategic authoring application and platform going forward for building information modelling and the building industry' (Laiserin 2002).

As the software houses developed their programmes, one of the first major pushes outside their remit saw the AIA devote a whole conference to BIM and its impact on the profession in 2006. They gathered the researchers of the intervening years and presented a new pan-cultural view of changes in the architectural profession. The previous year Norman Strong chaired an Integrated Practice Strategic Working Group which led the charge for the adoption of integrated practice as the primary emerging issue for the AIA board in 2006 (Broshar *et al.* 2006).

Notably, they wrote:

> Imagine a world where all communications throughout the process are clear, concise, open, transparent, and trusting; where designers have full understanding of the ramifications of their decisions at the time the decisions are made; where facilities managers, end users, contractors and suppliers are all involved at the start of the design process; where processes are outcome driven and decisions are not made solely on first cost basis; where risk and reward are value-based, appropriately balanced among all team members over the life of a project; and where the profession delivers higher quality design that is sustainable and responsive. This is the future perfect vision of Integrated Practice.
>
> (Broshar *et al.* 2006: 3)

The AIA pointed to the shift from a linear perspective to virtual modelling and they sought to define its impact on the relationship between the logic of representation and the logic of construction. Thom Mayne (2006) was even more forthright, or damning, with his paper 'Change or Perish' where he starkly stated to the delegates to prepare for a profession that they would not recognise in ten years' time.

Eastman pointed to two enabling technologies, parametric modelling and building data models, in *University and Industry Research in Support of BIM* (Eastman 2006). Significantly, Jonassen, in *Changing Business Models in BIM-driven Integrated Practice*, foresaw the need for a new role: model manager (Jonassen 2006). Jonassen addressed owner discontent, cost overruns and improvements in building information modelling. He saw a new approach to building delivery (note: the term 'integrated practice' was to become 'integrated project delivery') integrating design and construction, while entertaining enterprise operations to improve quality, productivity and safety, and reduce the cost and time of project delivery.

In *Roadmap for Integration*, the ever-increasing consumption of water, raw materials, fossil fuels and other non-renewables was tackled as eroding the global environment faster than it could be replenished (Lesniewski *et al.* 2006). In 'Suggestions for an Integrative Education', a call was made for radically changing education 'to shape the trajectory of exploration after graduation' (Lesniewski *et al.* 2006: 1).

Kimon Onuma painted a picture of the architect in the twenty-first century being transformed by process rather than software (Onuma 2006) and so improving value to clients in *The Twenty-first Century Practitioner*. Joseph Burns saw standardisation opening the door to allow analysis of designs at an early stage, reducing rework and enabling sharing (Burns 2006).

Technology, Process, Improvement, and Culture Change stressed BIM is not a new drafting tool (Bedrick and Rinella 2006), and noted the need for cultural change. *International Developments* drew attention to the global nature of architecture (Howell 2006), whether it is sourcing materials or working across borders, either physically (with worldwide offices) or otherwise (through global commissions).

Finally, life-cycle assessment is dealt with in *Information for the Facility Life Cycle* by establishing an overall facility life-cycle information strategy as well as methods to determine the handover, develop the handover and implement the handover (Fallon and Hagan 2006) in a way that uses the metadata that is now being generated.

These ten papers plus the introduction were to drive the AIA's focus substantially in the future; the following year the terminology changed or developed slightly, giving us today what is called IPD. The AIA's 2007 conference in California began addressing methods with 'what, why and how' and great stress was placed on achieving IPD (Eckblad *et al.* 2007).

References

Alexander, C. (1964) *Notes on the Synthesis of Form*. New edn (1 July 1974). USA: Harvard University Press.

Bedrick, J. and Rinella, T. (2006) *Technology, Process, Improvement and Culture Change: Report on Integrated Practice*. USA: AIA.

Belardi, P. (2014) *Why Architects Still Draw*. USA: MIT.

Broshar, M., Strong, N. and Friedman, D.S. (2006) *Report on Integrated Practice*. USA: AIA.

Burns, J.G. (2006) *Applications in Engineering: Report on Integrated Practice*. USA: AIA.

Eastman, C. (2006) *University and Industrial Research in Support of BIM: Report on Integrated Practice*. USA: AIA.

Eastman, C., Teicholz, P., Sacks, R. and Liston, K. (2008) *BIM Handbook: A Guide to Building Information Modeling: For Owners, Managers, Designers, Engineers, and Contractors*. USA: John Wiley & Sons.

Eckblad, S., Rubel, Z. and Bedrick, J. (2007) *Integrated Project Delivery: What, Why and How*, 2 May 2007, AIA.

Egan, J. (1998) *Rethinking Construction: The Report of the Construction Task Force*. UK: Department of the Environment, Trans Construction Task Force.

Elvin, G. (2007) *Integrated Practice in Architecture: Mastering Design-Build, Fast Tract, and Building Information Modeling*. 1st edn. USA: John Wiley & Sons.

Emmitt, S. (2007) *Design Management for Architects*. Oxford: Blackwell.

Emmitt, S., Prins, M. and Den Otter, A. (2009) *Architectural Management*. 1st edn. UK: John Wiley & Sons.

Erlanger, S. and Gohin, M. (2011) *Tycoon's Project: Nimby With a French Accent*. New York: Times Europe.

Fallon, K.K. and Hagan, S.R. (2006) *Information for the Facility Life Cycle: Report on Integrated Practice*. USA: AIA.

Ferroussat, D. (2008) *The Terminal 5 Project – Heathrow*. UK: BAA Heathrow.

Gehry, F. (2008) Digital Project – Frank Gehry, 26 November 2008.

Hardin, B. (2009) *BIM and Construction Management: Proven Tools, Methods and Workflows*. Indianapolis, IN: Wiley Publishing.

Haste, N. (2002) *Terminal Five Agreement: The Delivery Team Handbook* (without PEP). Supply Chain Handbook. London: BAA.

Howell, I. (2006) *International Developments: Report on Integrated Practice*. USA: AIA.

Jernigan, F. (2007) *BIG BIM little bim*. 2nd edn. USA: 4 Site Press.

Jonassen, J. (2006) *Report on Integrated Practice*. USA: AIA.

Kristensen, E.K. (2011) Systemic Barriers to a Future Transformation of the Building Industry From a Buyer Controlled to a Seller Driven Industry: An Analysis of Key Systemic Variables in the Building Industry, such as 'Procurement Model', 'Buyer Perception', 'Production Model', and 'Leadership and Management', Principally in a Danish Development Context and Seen From the Perspective of the Architect. PhD. Aberdeen: Robert Gordon University.

Krygiel, E. and Nies, B. (2008) *Green BIM: Successful Sustainable Design with Building Information Modeling*. USA: John Wiley & Sons.

Laiserin, J. (2002) 'RevitDesk versus AutoRevit', *The LaiserinLetter* (tm) Issue 11, 12 August, [online]. www.laiserin.com/features/issue11/feature02.php [accessed 31 October 2010].

Latham, M. (1994) *Constructing the Team: The Final Report of the Government/Industry Review of Procurement and Contractual Arrangements in the UK Construction Industry*. London: HMSO.

Lemoine, B. (2006) *Gustave Eiffel: The Eiffel Tower*. Hamburg: Taschen GmBH.

Lesniewski, L., Krygiel, E. and Berkebile, B. (2006) *Roadmap for Integration: Report on Integrated Practice*. USA: AIA.

Lion, R. (2004) 'Terminal 5 Single Model Environment – Vision, Reality and Results', *Excitech Design Productivity Journal*, 3(3), pp. 53–58.

Loos, A. (1930) 'I Do Not Draw Plans, Facades or Sections', Adolf Loos and the Villa Müller, [online]. http://socks-studio.com/2014/03/03/i-do-not-draw-plans-facades-or-sections-adolf-loos-and-the-villa-muller/ [accessed 31 December 2014].

Mack, P.E. (2005) Engineering Education in the 19th Century [Homepage of Clemson University], 9 February, [online]. http://virtual.clemson.edu/caah/history/FacultyPages/PamMack/lec122/eng19.htm [accessed 26 February 2013].

Mayne, T. (2006) *Change or Perish: Report on Integrated Practice*. USA: AIA.

Onuma, K.G. (2006) *The Twenty-First Century Practitioner: Report on Integrated Practice*. USA: AIA.

Potts, K. (2009) Project Management and the Changing Nature of the Quantity Surveying Profession – Heathrow Terminal 5 case study [Homepage of Royal Institution of Chartered Surveyors], [online]. www.rics.org/site/scripts/download_info.aspx?fileID=3099&categoryID=564 [accessed 31 December 2009].

Scheer, D.R. (2014) *The Death of Drawing: Architecture in the Age of Simulation*. London: Routledge.

Sigurðsson, S.A. (2009) Benefits of Building Information Modeling. Bachelor of Architectural Technology and Construction Management. Copenhagen: Copenhagen School of Design & Technology.

Smith, D.K. and Tardif, M. (2009) *Building Information Modeling: A Strategic Implementation Guide for Architects, Engineers, Constructors and Real Estate Asset Managers*. Hoboken, NJ: John Wiley & Sons Inc.

Smyth, H. and Pryke, S. (2006) *The Management of Complex Projects: A Relationship Approach*. Oxford: Blackwell.

Smyth, H. and Pryke, S. (2008) *Collaborative Relationships in Construction: Developing Frameworks and Networks*. London: Wiley Blackwell.

Vasari, G. (1991) *The Lives of the Artists*. Translated with an introduction and notes by J.C. and P. Bondanella. Oxford: Oxford University Press (Oxford World's Classics).

Wallas, G. (1921) *Creative-Process*, [online]. www.eclecthink.com/wp/wp-content/uploads/2013/11/Creative-Process-largeprint01.jpg [accessed 3 March 2015].

Winch, G. (2002) *Managing Construction Projects: An Information Processing Approach*. Oxford: Blackwell Science.

4 Culture shifts
Evangelical and evolutionary models
for BIM uptake

4.1 The roles of Government and its agencies

The Cabinet Office, in mandating BIM strategy for all public projects from 2016, set out four key tenets. First and foremost, they want 33 per cent lower costs, 50 per cent faster delivery, 50 per cent lower emissions and 50 per cent improvements in exports (Calvert 2013).

A roundtable working group hoped to achieve 20 per cent reduction in building costs (and promoted a 'buy four, get one free' catchphrase); that 33 per cent reduction in costs could be gained over the lifetime of the buildings; a 47–65 per cent reduction in conflicts and rework during construction; a 44–59 per cent increase in the overall project quality; a 35–43 per cent reduction in risk and better predictability of outcomes; 34–40 per cent better performing completed infrastructures; and a 32–38 per cent improvement in review and approval cycles (Waterhouse *et al.* 2011). Paul Morrell, the Government's Chief Construction Adviser of that strategy, also supported and encouraged the RIBA New Plan of Work.

A plan of work is an agreed documented instalment plan, which allows parties to have a method of paid milestones, beneficial and understandable to all stakeholders. The parties set out what is expected in return for what will be paid for each package of work (RIBA and ARUP 2013). As a method, they demand that the work is not undersold or oversold. To be undersold means that the design team are claiming money for incomplete or incompetent work. Oversold, risks the client severing the project early, having received more than for what they are paying.

With the onset of BIM, often there is more work done earlier in the project than is being remunerated for each work stage. Taken together with soft landings, if there is handover of the model, the design team is at risk. This mindset is the result of the litigious, distrusting methods of yesteryear. Currently, it is not unheard of that the models are rendered unusable, or that worksets or other parcels of work are disabled or even removed. These tasks can take more time and eat more of the budget than the original work itself, and are clearly not sustainable.

Fees and workload need to be better aligned to alleviate these concerns. As a trend, this implies more fees upfront, which banks and funding institutions are loath to commit to so early. Two things emerge from this juncture: first, one is that 'effort must be properly rewarded' and second, better trust needs to be nurtured throughout the sector, where an 'I will not sue you' mentality prevails (Smith and Tardif 2009).

Consensus contracts are appearing more and more to avoid this situation (ConsensusDOCS 2011). Levels of Detail and Levels of Information with proper filtering will also aid the matter (NBS 2015). This will allow the pertinent information to be delivered to the respective stakeholders. Ultimately, clients/owners or users assimilate or consume this data (it is what they are commissioning and rightfully expecting), so the industry needs to step up to the plate.

- 0 Strategic Definition

This is essentially a new work stage, where there is scope to validate the client's financial situation. Is the project sustainable, is it needed, what is needed and is it the best use of the client's resources? The core objective of the Strategic Definition is to identify the client's business case and brief, so that the project is founded on a sound sustainable basis. The project team can be discussed and assembled either in part or in full, so that a programme can be established (Ostime 2013) to set the project in motion. This is where the strategy can be checked, challenged or changed as befitting the project's scope, impact or commitment.

- 1 Preparation and Brief

The next stage kicks off the project into realisation, with a structure being put in place. In Preparation and Brief the project can be developed with clear objectives, covering quality outcomes, aspirations and budget. As Briefmaker, feasibility studies and site data can be quantified and set up. Roles and contractual obligations, including handover and manageable risks, can be categorised, and the process can be reviewed with regard to services, design responsibilities, information technology exchanges, communication methods and standards, all of which can be addressed in collaboration (Ostime 2013). There is also scope for a shift from the design team to the project team (more contractual relationships) to be discussed, affirmed and agreed upon.

- 2 Concept Design

As the Concept Design begins, outline proposals are made, including for structural design, extent and degree of services, outline specifications and preliminary cost information. Strategies for the project's procurement allow a final project brief to be issued (Ostime 2013). This includes levels of sustainability and life-cycle analysis regarding operations and maintenance. Evidence-based Design (EBD) issues can be raised here and acted upon. Co-ordination of the project now begins in earnest.

- 3 Developed Design

Here the concept is further developed by being co-ordinated with the other stakeholders' input, where a structural design is now proposed, together with an appropriate service system.

Figure 4.1 Stærevej 30–36 Algae Bio Reactor project

Outline specifications and building-part cost estimates now should tally with the design strategy and programme.

- 4 Technical Design

All architectural, structural and building services information should now be co-ordinated following the design programme. Information exchanges might vary depending on the selected procurement route, but should include specialist subcontractor designs and specifications.

- 5 Construction

The construction programme will be qualified by the subsequent design queries if and when they arise. On-site and off-site works are co-ordinated and include regular site inspections together with reviews of progress. Health and safety is also updated and closely administered.

- 6 Handover

Handover brings closure to the building contract and concludes the contract. Strategies carried out here include channelling feedback about the future life of this facility or future ones.

- 7 In Use

In Use heralds the implementation of the handover strategy, any post-occupancy evaluations and a review of the project performance, outcomes and any research and development that

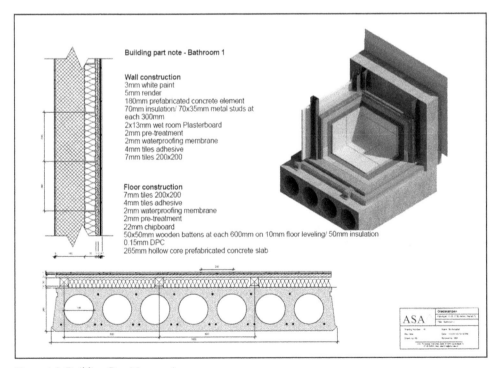

Figure 4.2 Building Part Note project

might have accrued under way. As-built information should be updated from ongoing client feedback and any operational and maintenance developments that may arise.

4.2 Professional bodies and umbrella organisations in the UK

PAS 1192-2:2013 is a specification for information management for the capital/delivery phase of construction projects using building information modelling (BSI 2014). It could be seen as the management layer laid over the skeletal RIBA Plan of Work. While the Plan of Work is a very topical subject, it is only a well guessed assessment of how new work practices are mapped (RIBA and ARUP 2013). There is a pressing need for continual critique and enhancement of the plan in a number of BIM-related areas, including soft landings.

The plan is a blueprint for a longer engagement between all the stakeholders in the process of constructing buildings (Sinclair 2013), as demonstrated in the new work stage 'In Use'. Its adoption of BIM sets out a course for moving away from antiquated pre-industrial methods (Kristensen 2011) to a more streamlined postmodern process where elements, assemblies and mass production enter the fray, becoming the bedrock for future work (MacLeamy 2010). Digitalisation opens the door for a matrix of analysis, performance metrics and simulation to take centre stage, instead of the trusted pencil and tracing paper, which cannot be tested.

PAS literally means a Publically Available Specification, referring to the standard methodology for managing the production, distribution and quality of construction information within the construction sector, by and large looking across the whole supply chain (BSI 2014). What must be asked then: are soft landings delivering this overarching aspect in the supply chain through to facility management or aftercare/user involvement?

Finally, while the subject deals with developing issues related to design management and how the profession deals with the integration of new management systems, it is imperative to stress that the soft landings encompass the life-cycle analysis, while a second objective is to assess the impact this has on embodied CO_2 while sensoring/monitoring the building's post occupancy. Architectural technologists stand to benefit most in delivering this critical piece of the jigsaw, working in unison with, but complementary to, architects and the other stakeholders involved (Barrett 2011).

The new plan paints a welcome broader picture of how procurement is only a part of a building's life, bringing initial strategies and sustainable mindsets to the fore (RIBA and ARUP 2013). It extends, gladly, this process into the everyday use of the building, with post-occupancy checklists and environmental performance data, while ultimately giving society a better end product. This hopefully then restores certainty and replaces rhetoric with reality (Harty and Laing 2010).

It is widely accepted that making a building is a process. It goes without saying that you cannot buy a house off a shelf. No two buildings share the same site simultaneously and hence there are basic differences, such as location and orientation, just to set the scene. Granted, there are typologies and catalogues, but essentially every building is a new beginning. Not only are the physical buildings different, but the planning and the procurement involving differing parties, and their occupancy, reflect differing peoples and their cultures.

This process is a collaboration between very many stakeholders over time, expressed through phases. These phases, or work stages, are designed to establish tangible milestones through the process, to aid the exchange of predefined work packages in return for remuneration. Increasingly, these packages are being addressed as soft landings (BSRIA 2014), after each phase, in a bid to better inform the client and to ease the process, ensuring that all on board are up to date and best informed.

Integrated Project Delivery requires a professional handling to administer it above and beyond a mere contractual document, to define the parameters' scope and to understand each stakeholder's value. Filtering who gets what does not happen automatically; managing the enterprise needs someone who knows how to dovetail with other professionals. This situation is best suited to a hands-on technologist who can leverage things, and has an all-round basis in construction (Barrett 2011).

4.3 Interactions between educators and practice

Educational success is no longer about reproducing content knowledge, but about extrapolating from what we know and applying that knowledge to novel situations.

(Schleicher 2010)

The relevance and interest of all the above to this work is that BIM is transforming the whole approach. While the input and output are essentially the same, everything in between is in a state of flux. But to its credit the framework is still holding and the experiential learning pedagogy is well placed to adopt this new era. The new buzzwords in connection to education are 'collaboration' and '(modular) outcomes', which are cornerstones to the teaching methods.

What is missing are upskilling and lifelong learning, which are fast becoming the new goals of the twenty-first century. Teaching by rote does not equip the student of today for the changing world of tomorrow (Schleicher 2010). The models being developed worldwide for higher education all have an important element, which includes a section on lifelong learning and career change studies. If we do not follow suit, the workforce will not be as adaptable to technological and social change (Kelleher 2009).

This is very relevant for the construction industry. While industry and its clients share a range of strategic interests with higher education, which include a stream of suitably trained graduates, currently there is little or no Continuing Professional Development (CPD) for existing personnel, research with international market potential or access to beneficial collaborative partnerships. New mechanisms such as postgraduate programmes might allow learners to develop a portfolio of modules which lead to a particular or specialist qualification, but do

Figure 4.3 Lighting Calculation Analysis

not necessarily come from the same institution, discipline or mode of delivery. There is also a chance to exploit the unique benefits of IT to increase access and participation.

The introduction of a policy and funding framework based on transferable credit-based learning that came with the Bologna Directive in 1999 (EHEA 2011) would help tailor new modes of delivery and a provision to enable greater access opportunities. Increasingly, flexibility of provision will be a key indicator of the responsiveness to society. The Bologna Process can also facilitate a substantial reorientation in programme delivery towards part-time and flexible courses.

There is also a general consensus that a strong relationship between business and higher education is critical to economic competitiveness. Business even favours the concept of becoming a real partner and voice within higher education. Stronger partnerships with the business community, including with SMEs, bring opportunities for educational institutions to enhance the prospects for graduates. Therefore, higher education institutions need to respond better and faster to the demands of the market and to develop partnerships with businesses and others, which harness scientific and technological knowledge. The potential of higher education institutions to contribute to the economic, social and cultural development of their regions is far from being fully realised and cannot be underestimated.

Figure 4.4 Model of scenery for the Sala Nova Aldo Rossi, 1997

Figure 4.5 Vicenza Basilica Palladana

Figure 4.6 Autodesk Revit model of Vicenza Basilica Palladana

Higher education institutions must become more entrepreneurial, widen their service portfolio and address the needs of a wider range of organisations and employers. They should be more dynamic in their approach to collaboration. If industry can be brought in as a key stakeholder, as it must beyond the impact that it has today, then new avenues need to be opened up in the curriculum and new courses offered to meet the demand.

4.4 Smoothing out the curve for SMEs

While soft landings try and address underperformance and disappointment, many post-handover problems can be caused by inadequate preparation. By introducing a concept of aftercare, a situation is set up which allows the design team to remain in the loop, and the users to address shortfalls. Ultimately, with constructive feedback, the 'three pillars' of constructing excellence can be met: Post-Occupancy Evaluation (POE), true EBD and better project briefing. EBD is the realisation that design options should be based on researched evidence rather than the intuition of the designer (Ulrich 2001). This is relatively new in construction, as research is in its infancy, unlike pharmacy and medicine.

As said, soft landings provide professional aftercare because too many buildings have been put into service in the past without being fully commissioned (BSRIA 2014). The entire process gets adopted, meaning that risks and rewards are shared. Soft landings provide leadership, bringing trust and respect to all stakeholders. They set roles and responsibilities so that the client is an active participant, which ensures continuity. This commitment to aftercare not only reviews and fine-tunes performance for the project in question, but also raises awareness across the board. Incentives used to deliver the facility encourage higher performance, while the feedback informs both current and future design. They also bring operational outcomes into focus, which is a reality check against requirements and ambitions.

Notwithstanding that, they also begin a dialogue with the building managers so that there is meaningful handover. Prospective occupants should be actively researched to understand their needs and expectations by setting out performance objectives in line with key performance indicators. Finally, regardless of the contractual obligations, team members need to be comfortable communicating with each other, irrespective of who they are, be it a user specialist, subcontractor, or whatever (BSRIA 2014).

References

Barrett, N. (2011) The Rise of a Profession Within a Profession: The Development of the Architectural Technology Discipline Within the Profession of Architecture. OpenAIR@RGU, [online]. http://openair.rgu.ac.uk [accessed 13 August 2015].

BSI (2014) PAS 1192-2:2013 Specification for Information Management for the Capital/Delivery Phase of Construction Projects Using Building Information Modelling [Homepage of BSI], March, [online]. http://shop.bsigroup.com/Navigate-by/PAS/PAS-1192–22013/ [accessed 9 March 2014].

BSRIA (2014) The Soft Landings – Core Principles [Homepage of BSRIA], March, [online]. www.bsria.co.uk [accessed 4 March 2014].

Calvert, N. (2013) Why We Care about BIM… [Homepage of Directions Magazine], 11 December, [online]. www.directionsmag.com/entry/why-we-care-about-bim/368436 [accessed 28 March 2015].

ConsensusDOCS (2011) Why ConsensusDOCS [Homepage of ConsensusDOCS], [online]. www.refworks.com/rwbookmark/bookmarklanding.asp [accessed 9 April 2011].

EHEA (2011) European Higher Education Area website 2010–2020 | EHEA, [online]. www.ehea.info/ [accessed 13 January 2011].

Harty, J. and Laing, R. (2010) 'The Management of Sharing, Integrating, Tracking and Maintaining Data Sets, Is a New and Rather Complex Task', 2010, 14th International Conference of Information Visualisation (IV 2010), p. 610.

Kelleher, P. (2009) *Smyth Report: Thematic Synthesis. Written Submissions to Strategy Group.* National Strategy for Higher Education in Ireland.

Kristensen, E.K. (2011) Systemic Barriers to a Future Transformation of the Building Industry from a Buyer Controlled to a Seller Driven Industry: An Analysis of Key Systemic Variables in the Building Industry, Such as 'Procurement Model', 'Buyer Perception', 'Production Model', And 'Leadership and Management', Principally in a Danish Development Context and Seen From the Perspective of the Architect. PhD. Aberdeen: Robert Gordon University.

MacLeamy, P. (2010) The Future of the Building Industry (3/5): The Effort Curve [Homepage of HoK], February, [online]. www.youtube.com/watch?v=9bUlBYc_Gl4 [accessed 10 September 2011].

NBS (2015) Digital BIM toolkit [Homepage of NBS], 2 January, [online]. www.thenbs.com/bimtoolkit/bimtoolkit.asp [accessed 2 January 2015].

Ostime, N. (2013) *RIBA Job Book.* 9th edn. London: RIBA Publishing.

RIBA and ARUP (2013) *Designing with Data: Shaping our Future Cities.* London: RIBA ARUP.

Schleicher, A. (2010) *The Case for 21st-century Learning.* Paris: OECD.

Sinclair, D. (2013) *Assembling a Collaborative Project Team: Practical Tools Including Multidisciplinary Schedules of Services.* London: RIBA Publishing.

Smith, D.K. and Tardif, M. (2009) *Building Information Modeling: A Strategic Implementation Guide for Architects, Engineers, Constructors and Real Estate Asset Managers.* Hoboken, NJ: John Wiley & Sons Inc.

Ulrich, R.S. (2001) Effects of Healthcare Environmental Design on Medical Outcomes, Design and Health: Proceedings of the Second International Conference on Health and Design. Stockholm, Sweden: Svensk Byggtjanst 2001, pp. 49–59.

Waterhouse, R., Morrell, P., Hamil, S., Dr, Collard, S., King, A., Clark, N., Kell, A. and Klaschka, R. (2011) BIM Roundtable Discussion [Homepage of NBS], 13 April, [online]. www.thenbs.com/roundtable/index.asp [accessed 10 September 2011].

Part II

A contemporary view of BIM in the UK

5 Drivers for innovation

Figure 5.1 Halmlageret, Carlsberg, ReCal capture

5.1 The Internet changed everything

Practically speaking, augmented reality (AR) is the layering of a virtual world on to a real one. It allows both worlds to be experienced together at the same time, meaning a proposal can be assessed before being realised. Where this will truly pay dividends is when the model filters down to the least expected actor in the project.

This can be achieved using a handheld device or a device worn like glasses with heads-up data on the fly. A holographic headset has been announced for release with Microsoft Windows 10, which can superimpose a totally immersive world into the viewer's cone of vision (Kelion 2015). A TV screen can be imposed on a wall, directions for fixing plumbing, or carrying out other household tasks can be made at the user's behest, and prototypes in design can be shown in real time for others to comment or reshape in a collaborative environment. This means that design implications can be tested at design team meetings, dynamically, with all the relevant stakeholders present to comment on the alternative design proposals. This is known

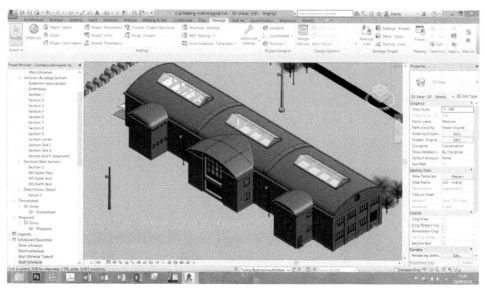

Figure 5.2 Halmlageret, Carlsberg, completed 3D model

as Integrated Concurrent Engineering (ICE). This can be very important in clash/collision detection, but also at conceptual design phases, because it brings everyone up to speed.

AR on a building site could mean an electrician is able to see where a light switch is to be placed or mounted on a wall without the need for drawings or specifications; instead all is at hand on a mobile device. Cabling tracks can be superimposed on the wall, along with the demarcation of other utilities like water pipes, so that costly mistakes are avoided. Text or voiceover can indicate what is to be executed at the user's pace. Positive feedback can be given when the task is correctly completed, with a sign-off and release of funds, rather than the retention of monies in case of litigation.

RFID (radio-frequency identification) chips can sequence what is to be filtered so that there is no data overload. They activate when they detect a host to which they are programmed to respond. Typically, suppliers embed them in their products and they activate when a device is sensed nearby, for example a precast concrete flight of stairs can have fixing instructions embedded near the corners of the component, or activate links to digital manuals flagged when being lowered into place. These can be viewed on handheld devices or retina screens as before.

Already, we are seeing wearable technologies, in the form of glasses (e.g. Google Glass) which can map the virtual world on to the real. These technologies can also address reports and documents and can answer and retrieve data when asked. This means that arriving on site, a request could be made for an update of the project's progress. This could be furnished through text with focus set to infinity on the retina of the lens scrolling with eye movement, or be spoken or read out as the recipient approaches hotspots while traversing through the site, or it could be pre-mapped with the revisions placed correctly in 3D space so that best fit, correct mounting and flawless integration can be checked and approved. Once repaired or updated, a status stamp can verify completion or approval.

Robust technologies in the form of headsets can immerse the user in the 3D world, giving a complete package. This has tremendous possibilities in convincing the client when marketing the project. The gaming industry (e.g. Oculus Rift) is bringing product to market that is quite remarkable (Brouchoud 2014). This also impacts urban planning officials, third-party interest groups and local residents.

Targets, such as 2D drawings, can host 3D models which respond to the viewing angle and position of the target. This improves real-time visualisation, cultivating collaboration and communication among stakeholders. The objects or models can have 4D attributes added so that exploded views can be shown or sequences explained while being viewed through a handheld device (Yoders 2014). Scaling can also occur, allowing a reduced size for demonstration purposes, or full scale for urban issues like placement, context and appropriate environments to be assessed. With markers, or with increasingly better GIS methods, the virtual world can be viewed and appreciated in its intended surroundings.

As sought-after in the Latham report, better health and safety has long been on the cards in the construction industry. Increasingly, with globalisation, and especially in Europe (due to tendering processes), building sites can be and often are multicultural. This can cause unforeseen problems if health and safety procedures are not understood across the board because of language difficulties. Hotspots can even flag up where danger approaches.

AR offers a method of tailoring the message in different languages either to each situation or specifically to the imminent danger faced. It remains up to date and has a single source of revision. Safety checklists are at hand and the latest safety meeting's minutes published directly after the meeting (Jones 2014).

AR will also increase speeds of execution on site with better implementation of BIM. There are also many new benefits with the advent of 'drone' technology, in surveying and capturing site geometry. This allows unmanned small crafts to fly around and through a building site capturing the terrain, existing buildings and existing context. Then through photometry or scanning (Autodesk Recall for example), it can be pieced together through alignment of common features in the photos or scans to give an accurate survey of the existing conditions.

This can then be choreographed into big-data models using Geographical Information Systems (GIS) to map the data for new purposes, but from the far end of the other spectrum. As the name suggests, it starts with the global geographical things and narrows down to elemental things from there. As agents of archive collect big-data, many possibilities emerge: trends and tendencies can form patterns, areas for concern can be identified for action on an urban level, and infrastructure can be best administered, allowing for the measuring of consequences.

In Oslo, beta testing is happening where building control can be digitally administered. This means that an International Foundation Classes (IFC) model is uploaded to the municipality's server, where code checking takes place to allow the proposals or grant conditional permits. Over a 20-year span, it is expected that the city will acquire much of the central area. As the city are holding agents of this data, number crunching can occur to check climatic impacts or assess energy loss over districts across the city, meaning that accountable decisions can be made on future expenditure (Rooth 2010).

GIS has now matured into a foundation technology that seamlessly provides 'world-to-the-widget' scalability. This means that it can both drill down to very small scale (from the larger geographic expanse) and also bring minute layering of data into the matrix. On the one hand, it can tell you how many unoccupied offices are within 500 meters of a parking space or how many employees will have to travel more than half an hour to get to an office location. On the other hand, it can intimately map an asset by building, floor, room – all the way down to the equipment and its usage in a Building Information Spatial Data Model (BISDM). This can involve sensory monitoring of the facility so that equipment can be observed in its use or non-use, giving qualified information. Sensors can collect vast amounts of data, which can be crunched through algorithms to offer incentives to curb consumption at peak times and increase demand at lower times (The Economist 2015).

5.2 Is there an inevitability about BIM as a collaborative medium?

Consider, first, the clause from AIA Document B141/CMa section 2.6.6 which states: 'The Architect shall not have control over or charge of and shall not be responsible for construction means, methods, techniques, sequences or procedures, or for safety precautions …' Compare this with the legal jargon that Frank Lloyd Wright (Bennets 2010) used, which stated:

> The architect undertakes to itemise mill work and material for the building, lets contracts for piece work and eliminates the general contractor where possible by sending a qualified apprentice of the Taliesin Fellowship at the proper time to take charge, do the necessary shopping and hold the whole building operation together, checking cost layouts, etc., and endeavour to bring this work to successful completion.

In 2011, the RIBA (2015) produced a report, *The Future for Architects*, where Dickon Robinson, Chair of Building Futures, said:

> Architects are not alone in needing to respond to the impact of a globalising economy, exploding information technology capability and cultural confusion. However in the face of a continuing erosion of traditional architectural skills to other players, the profession seems peculiarly vulnerable to a nostalgic backward glance at a bygone age in which the architect was the undisputed boss.

Among the other comments in the report were: 'Architects have shed project management, contract administration, and cost, and ultimately if they lose design co-ordination, then you have to ask what they are there for …', as well as 'I think the entire middle-sized practices from about 25–150 people will be gone, and we'll end up with two very distinct types of practice at each end of the market.'

The companies that have not implemented BIM will undoubtedly be left behind and will not be able to take on certain jobs where the builder has BIM as a demand of delivery. It will also make them less capable of holding the project prices down, compared to the BIM users who will benefit from that advantage. As Patrick MacLeamy says:

> It all begins with BIM; the architect uses 3-D modelling to investigate options and test building performance early on in order to optimise the building's design. The design is then handed off to the contractor who streamlines the building process with BAM (Building Assembly Modelling), which allows for a significant decrease in construction costs. Once complete, BAM is turned over to the owner and becomes BOOM (Building Owner Operator Model). This allows the owner to manage the building over time and ensure optimised building performance throughout its entire life cycle. The real promise of 'BIM, BAM, BOOM!' is 'better design, better construction, [and] better operation.
>
> (MacLeamy 2010)

The term 'BIM, BAM, BOOM!' explains the way we should use and think about the model through the project. Whereas BIM is related to building information and design development, BAM is related to IPD in the construction management phase, and BOOM is in the facility management phase.

Increasingly, large public clients are mandating the use of BIM in open formats. Aconex Limited (ASX: ACX) is a global provider, based in Australia, of mobile and web-based collaboration technologies. It offers a cloud-based Software as a Service (SaaS) solution for project information and process management. In 2009, Aconex managed fewer than 10,000 models. By the start of 2013, it was managing over 270,000 models – a growth of 2,600 per cent in just three years. The model has also been getting more detailed, and this increases its size. In the last five years, the average models have more than doubled in size to 54 Mb (Aconex 2015).

Meanwhile, we are placing ever greater demands on our built environments and by adding multiple components to our buildings, we increase the probability of errors. This can be policed with rollback back ups, especially in cloud environments. The ability to just delete or change something in the model is also much easier than with paper drawings. The model allows better control over:

*   buildability;
*   progress;
*   access to solutions; and
*   project economy.

Typically, a project manager might hold fortnightly project review meetings. This might occur on Thursdays (on even weeks). This would require delivering updated models Tuesdays, (on even weeks). BIM co-ordinators then would assemble federated models and begin collision reports to be ready for the review.

ICE is a relatively new design management system that has had the opportunity to mature in recent years to become a well-defined systems approach towards optimising engineering design cycles. The basic premise for concurrent engineering revolves around two concepts.

The first is the idea that all elements of a product's life cycle, from functionality, producibility, assembly, testability, maintenance issues, environmental impact and, finally, disposal and recycling, should be taken into careful consideration in the early design phases.

The second concept is that the preceding design activities should all be occurring at the same time, i.e. concurrently. This way, errors and redesigns can be discovered early in the design process when the project is still flexible. By locating and fixing these issues early, the design team can avoid what often become costly errors as the project moves to more complicated computational models and eventually into the actual manufacturing of hardware.

In practice, this can mean all technical stakeholders at a design meeting might be wielding hand-held devices as they address in real time issues normally noted and taken home from meetings to be rectified by house staff and presented at the next meeting.

5.3 Evangelical and evolutionary paradigms for BIM uptake

Effective collaboration is fast becoming the clarion call within the building industry. It is closely related to building information modelling. Where the two meet is a blend of management and application. BIM is a digital process that builds a virtual model of the real world and in so doing allows analysis, simulations and full testing before the foundations are even marked out, as well as creating a useful platform for communicating with clients, authorities and financial stakeholders.

Traditionally, many of the problems within construction only come to light on site, leading to delays, variations and counterclaims, which in turn lead to buildings being costlier and late, promoting client dissatisfaction. For clashes and collisions of building components to be resolved in the design phase, where it is easier to rectify and cheaper to remedy, requires all stakeholders to be involved earlier in the whole procurement.

Authoring the model with floors, walls and roofs, as well as doors, windows and all the other components that go to make the building, allows for photo realism, fly-throughs and immersive environments. The model can be analysed for thermal, daylighting and solar performance, and quantified with material take offs, costed with indexed pricing and sequenced for optimal construction phasing. This means certainty for clients, control for architects and code checking for planning authorities.

Such a ubiquitous war chest should have the industry clamouring for its adoption, but sadly not! Entrenched methods and the old guard disciplines are reluctant to share data, pool resources or even talk to each other without a contract and lawyer present. 'Architects giveth, engineers taketh away' and 'engineers think architects are dreamers, and architects think engineers are killjoys', are just two examples of the prejudices that litter the professions

By contrast, contractor organisations are beginning to see the benefits of such systems and are advising their members to build a model before tendering for work if it is not already part of the package. A while ago, the debate was about 'if, why and how' the model might be given to contractors. Today, contractors are mandating models in any way, shape or form as it provides some sort of flow across the process; needless to say, the better the model the better the flow. If there are high demands for facilities management information, and if there is not, why not, then the debate now progresses to how to deliver the model to the client/user. Here, there is a translation need from objects to assets.

5.4 Industry and Government push-pull initiatives: reviewing cultural and business models for practice

Generally, green building projects are increasingly popular, as they become viewed as long-term business opportunities. Architects, engineers, contractors and owners (AECOs) are reporting that 60 per cent of their projects are green, up from 13 per cent in 2009 (Bernstein and Mandyck 2013). But in order to claim a sustainable mandate, there must be a measurable matrix into which the numbers can be verified and banked. If you cannot measure it, you cannot deal with it. Sustainability comprises of three pillars: society, the environment and economy. Reyner Banham, introducing the reader to the machine age in the 1950s, comments on:

> our accession to almost unlimited supplies of energy ... to dispense some previously inconceivable product, such as an aerosol shaving cream, from an equally unprecedented pressurised container, and accept with equanimity the fact that he can afford to throw away, regularly, cutting-edges that previous generations would have nursed for years.
>
> (Banham 1970: 9)

Clearly, we are paying for these excesses today.

Elemental costs, for the economical element of sustainability, start out with calculating the unit price of a material. This is then combined with a cost factor for the assembled component, where the total cost can be shown, including the labour cost, material cost, hire of external equipment, soft costs, fees, and even a cost index for where the work is being realised. Labour and material costs are calculated with an exponential value so that when the volume of the work increases, the price per square metre decreases.

Similarly, other properties allow the calculation of u-values of the component, using the heat transmission of the material. Even the density of a material can be used for load calculations, meaning the whole process of 'information searching' can be shortened. Even fire and acoustic information can be just at your fingertips. The material can be linked to a data catalogue from a supplier which can include transport, carbon footprint, upcycling or recycling. Temperature line-loss across a construction can be mapped and daylighting factors plotted across floor plans.

Put another way, in most cases the information can be just embedded in the material, where it has been verified by an appropriate institute or supplier. The most reassuring thing about the material library is that there is no information loss in any part of the process and the component description and the component itself match, making the project impeccable. This is a prime Government push initiative that needs to be handled sooner rather than later.

When constructing a building, it can be said that there are two main physical resources involved: the materials necessary to form the various parts, and the technical ability to assemble the parts into an enclosure (Osbourn 1985). The raw materials for a building are rarely used in their basic form, but are treated or processed in some way to suit their purpose (Stroud Foster 1979). They become the elemental pieces of the building and together with functional requirements and performance criteria form components, which can be deployed in the project. Renzo Piano, as a lifelong sailor and designer of yachts, used his a priori approach to explore different materials, ranging from marine plywood to ferro cement (Frampton 1997). But this is an iconic individual, exhibiting a large element of accumulated experience.

When looking at the performance of buildings, the demands placed on building components can be said to include load-bearing capacity, climatic envelope, insulating qualities, fire ratings, moisture control, ventilation, wind stopping, maintenance and acoustics. This list is not exclusive or comprehensive, but provides an insight in to some of the requirements to be met. Others could include aesthetics, cost, quality, ease of manufacture and erection.

A technologist's role is to affect this choice in the best manner possible (Emmitt 2012). Traditionally this was accomplished with a body of knowledge built up by experience from tried and tested methods. It led to conservative decisions being made in component development and led to building failures where innovative methods were misunderstood with dire consequences. In 1960, there were approximately 50,000 building materials. By 2000, this had trebled and it is ever increasing (Palsbo and Harty 2013). Expecting to acquire knowledge for all products out there is no longer feasible, but methods for finding the correct performance are needed.

Previously, materials like common brick or similar masonry could make a brave attempt at satisfying all or most of the requirements listed above, whereas modern assemblies invariably use sandwich constructions which comprise several materials all doing individual jobs from the list (Selck 1974), so that one component is load-bearing, another is waterproof, and another provides decorative surfaces, as in cladding for example. They therefore need to exist together with correct sequencing and tolerances, allowing them to breathe and perform in unison.

A second point that can be noted from this is that accurate methods of measuring that performance are relatively recent, suggesting that the demands made on the building stock were not robust or comprehensive previously. Being unable to measure something means that it cannot be dealt with adequately or methodically.

The idea of BIM is to gather as much relevant information in one place so that it can then be accessed by the widest possible array of different programmes (Eastman 2012). This makes it possible for all parties involved to access the latest information. This creates a new level of collaboration where the loss of information is limited to an absolute minimum, thereby offering greater accuracy in estimation and avoidance of error, which results in saving time and costs through all stages of the project.

To gain all the benefits of BIM, all the stakeholders must work with the same fundamental mindsets that not only changes the way the information is accessed but also the way it is archived and how it is handled through all stages of the project (Bernstein 2013). This means BIM is not only a new technology but also a new way of working at a strategic level. The opportunities that BIM creates to store the information in the 3D model is a huge advantage when all the parties are working with BIM.

Having all the information in one place makes the model the primary tool for document generation, and moves away from the conventional parallel world involving word-processing and CAD drawings (Pittard 2012). It creates a workflow where specifications and reports (that are both time-consuming and hard to update if changes are made) are an integrated part of the model and are automatically updated as the project evolves.

To ensure the effective sharing of the information pool, BIM-related software must be compatible with the 3D model so the required information can effortlessly be transferred from model to calculation, simulation and analysis. So essentially, BIM combines the technologies that we work with on a day-to-day basis (Kennerley 2013).

The most often used expression of the level of BIM is 3D (Building Design in 3D with Quantities), 4D (3D + Time), 5D (4D + Cost), 6D (5D + Facilities Management) and 7D (6D + Physical Performance). These levels are obtained by using the information from the model in a variety of different software programmes combining the results. BIM improves both the quality of the work and the delivered product. This creates easy access to project time planning, budget estimates, energy consumption, solar studies, construction timeline animation, visualisation, animation, and so on, by having all the information accessible from one place.

One of the things that makes BIM so hard to implement on a full scale is if one or more of the stakeholders does not engage in BIM. All the information that is loaded into the model then needs to be extracted and filed into conventional documents and reports for the relevant party. This means that the information is now processed more than once, which is time-consuming and, to some extent, not properly integrated into the BIM model.

The importance of the information inside the BIM software then becomes less important than the documents extracted. The way BIM works today is that all the information loaded into the model comes from an external source, which is sometimes easier to place directly into the conventional documents rather than loading into the model and extracting it afterwards.

Another relevant point is information loss that can occur between software programmes. BIM works impressively by exporting information to other programmes, but the information obtained in the external programmes often cannot be imported back into the model. This creates problems that can in some cases become overlooked, and can result in huge errors. For example, the hatch pattern on a material's sectional view might be dropped or not supported when exporting from one programme to another (Pazlar and Turk 2008), or a level of detail might not be properly mapped between two programmes, meaning that on transfer back it has vanished and it reports as a void in the data.

BIM has many working tools, which are becoming more and more integrated into one overarching programme that is more open and easier to use. From when the project starts until the construction is completed, users load information from external sources into the model. But would it not be better to make the model itself into the main source of this information?

If the model becomes a container of all the information from the start of the project, the user can then select the relevant information from the project through a process of filtering.

All the selected information can easily be passed on to the parties that need it. There is no need to handle the same information more than once, reducing double work. If we accept that the main information which is loaded into the model is material information, by having all the material information in the model or a model library resolves the problem.

Materials play a key role in construction and by definition should also have a key role in BIM. Mentioning a particular material to a competent craftsman implies the scope of work to be done, by whom, for how long and at what cost. Bringing this level of detail to modelling can be accomplished by creating a comprehensive material library where all the information is embedded in the model, reducing the opportunity for human error. This is a good example of the dynamic, bi-directional proper use of BIM.

The same key information should be standard information in any BIM-related software. By creating or using a library, you are sure that all the information is therefore in the model, meaning there is no need to add additional (parallel) information, which would open up the opportunity of human error.

With a fully updated material library the benefits are huge at the start of the project (Autodesk). Instantly after drawing a foundation, there is the opportunity to create a schedule with all the information needed to construct it (Bowers 2012). Then the engineer can access it, and any changes made are automatically updated in the entire documentation. While the concept is being made, the architect has ballpark figures for cost calculations for every component that is made.

The process for compiling a personal price book, setting up an external price book such as Spons or Wessex, or making a new one is essentially the same, although the fastest way is to use a system with a database that is compatible with the preferred authoring programme (e.g. Autodesk Revit), thereby importing most of the material information. The biggest advantage with a material library is shown when it is combined with schedules, where all the required information for the different components is automatically updated when a material is chosen, thereby saving both time and effort.

For example, material trade scheduling works by having the differing trades' information of the materials, and a list with the salaries of the different trades for easy updating. When the exponential decreases or worked-times are calculated, the amount of worked hours is multiplied by the salary.

Scheduling in Revit together with material information can list and categorise a range of information required for a project. The opportunities for information export will include labour time that is shown in hours or days, the construction time for the different materials, the amount of workers and even specific salary for the labour time.

The cost shows the price of each material in the construction and a combined price for the entire component, including labour costs, material costs, hire of external equipment, soft costs, fees, and even a price index for where the work is placed (built-up/urban or remote/rural, for example). Labour and material costs are calculated with an exponential value so when the volume of the work increases, the cost per square metre decreases.

Detailed descriptions are often used on components to easily set up type tagging, but the information is typically entered manually and is specific for each individual component, consisting only of the specific information about the materials in the component. By having the information predefined in the detailed material libraries, there is a better chance that the component tag is automatically updated with the latest information. By using this workflow,

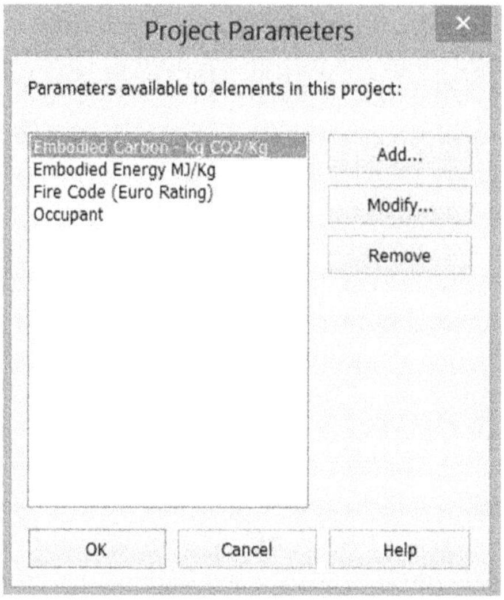

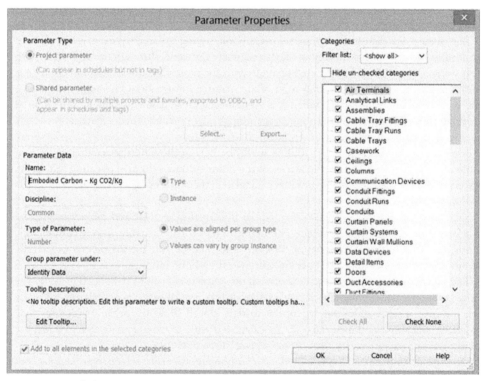

Figure 5.3 Autodesk Revit project parameters added

Figure 5.4 Autodesk Revit project parameters data

a description in standard constructions is made by a click of a button which assures the user that the information is accurate and always up to date.

The assembly description is information that is specific for each individual component or building part and to some extent does not involve the material library, but if the materials are detailed enough then it becomes a possibility. Sometimes a new material is related to another, similar, material in order to determine a trade and cost price. Of these materials, an assembly description is used which has an influence on the price and must be altered. Similarly, the assembly description can be added to all the materials in the library. The description is only limited by the level of detail required.

Trade information is usually restricted to the material's trade component. As the component is assembled with different materials, the different trades involved in the process of construction are shown in the component list, giving a clear overview of the work and the resources involved. This information can then be used for more detailed quantity take-offs and all the relevant information needed for a works specification.

In Autodesk Revit, there is a feature that can calculate the u-value of the component using the heat transmission of the material. Having the information directly in the material library ensures that the user has the correct information always and saves time, as the user does not have to search separately for an external document.

Likewise, the density of the material is often required for load calculation, so in the same way as with the u-value, the process of information searching is shortened. In the process of finding the right material for the component, material information like the span on the material according to sizes can help determine the total size of the component in the concept stage and thereby minimise the chances later in the project, where it can be more time-consuming. Even fire and sound information is right there at the end of your fingertips. So when the right material has been selected, the user is sure that aspects like fire resistance and sound attenuation are part of a better informed design decision.

In green projects, where the focus is on the carbon emissions of the construction and the carbon imprint of the materials, it can be a long process to calculate the carbon used to make the materials for the entire project, but having the information in the library the user will only have to set up a schedule that shows the entire calculation for the project. The schedule properties can also be saved as a template, so it becomes a standard calculation for all future projects, making designers think more about the environment and the use of materials. Schedules can also show life-cycle analysis of materials so if any of the materials far exceed the lifetime of the project, it creates the opportunity to look into the reuse of the material as in a cradle-to-cradle project.

Full lifetime operation and maintenance information for all of the materials can also be added to the material library, so the owner of the project can extract the information and know when to do what. Everything required to make an entire project specification, including health and safety information, can be placed in the material library, making it one of the most time-saving and secure tools that BIM can offer.

Like all other BIM-related processes, using a fully updated and well-maintained material library means the user must adopt a new mindset and workflow. The research for materials can no longer happen on the Internet or in the studio literature library, but in the model itself. The material can link to the data catalogue from the supplier, but in most cases the information lies in the material properties. The most reassuring part about a material library is that there is no information loss in any part of the process, and if the wrong material is selected by the user the component description, and the component match, flag this, making the project, once again, flawless.

The time-saving of a fully intergrated material library could be up to an estimated 10 per cent of the entire project, 40 per cent of the design stage and 80 per cent of the written documentation. Information handling is limited to specialists where the descriptions and performance information is divided between either in-house specialists or external sources that have experience in the subject.

Therefore the material library can be difficult and time-consuming to establish, so for small companies the investment in making the library might be expensive. The way that materials work in Revit at the moment also gives some minor complications, one of them being that two materials cannot be placed in the same layer.

As an example, insulation and rafters can occupy the same layering in a roof construction where the rafters are not physically placed but, mentioned in the description, it is possible to circumnavigate the problem. If there is a conflict, the solution is to place the elements as virtual components (studs/rafter, or whatever) in the insulation zone. They now feature in the model and can be extracted. Another method is to use the LODs in the model, so that the different elements can co-exist in differing levels.

The material library requires a lot of work to begin with, but when it is up and running it is a huge advantage both in the implementation of BIM and in all the subsequent work stages. Main suppliers of materials for construction have a great opportunity to make a material library in collaboration with their suppliers. This would be a huge advantage for BIM users, where the template could produce a shop-floor list with quantities that can be sent directly to the customer. Many suppliers are waking up to the potential of this product placement and are grasping this opportunity (BIMobject 2015; NBS 2015).

References

Aconex (2015) What is BIM? [online]. www.aconex.com/what-is-BIM [accessed 24 March 2015].

Autodesk (2012) Autodesk Revit: Managing Materials Libraries [Homepage of Autodesk], 11 May, [online]. www.youtube.com/watch?v=hSX7A2F26SQ [accessed February 2013].

Banham, R. (1970) *Theory and Design in the First Machine Age*. 4th edn. Glasgow: The Architectural Press.

Bennetts, R. (2010) Divide and Fall Source. *RIBA Journal*, 2010, October 2010, 10, 22.

Bernstein, H.M. and Mandyck, J. (2013) *World Green Building Trends SmartMarket Report (2013)*. USA: McGraw Hill Construction.

Bernstein, P. (2013) About BIM [Homepage of Autodesk], February, [online]. usa.autodesk.com/building-information-modeling/about-BIM/ [accessed February 2013].

BIM object (2015) Products | BIMobject [Homepage of BIMobject], 12 January, [online]. http://bimobject.com/product [accessed 13 January 2015].

Bowers, D. (2012) Sharing Autodesk Materials Throughout an Organization Applying Technology to Architecture [Homepage of DBC], 15 June, [online]. aectechtalk.wordpress.com/2012/06/15/sharing-autodesk-materials-throughout-an-organization/ [accessed February 2013].

Brouchoud, J. (2014) BIM Goes Virtual: Oculus Rift and Virtual Reality Take Architectural Visualization to the Next Level [Homepage of Arch Virtual], 19 January, [online]. http://archvirtual.com/2014/01/19/bim-goes-virtual-oculus-rift-and-virtual-reality-take-architectural-visualization-to-the-next-level/#.VLPqvyvF-Sp [accessed 13 January 2015].

Eastman, C. (2012) Building Information Modeling 'What is BIM?' [Homepage of Georgia Tech], 3 December, [online]. http://bim.arch.gatech.edu/?id=402 [accessed 3 December 2012].

Emmitt, S. (2012) *Architectural Technology*. 2nd edn. UK: Wiley-Blackwell.

Frampton, K. (1997) 'Renzo Piano: The Architect as Homo Fabien', in: *GA Architect, 14 Renzo Piano Building Workshop*. Japan: Y. Futagawa, pp. 8–28.

Jones, K. (2014) Five Ways AR Will Benefit the Construction Industry SmartReality [Homepage of Construction Data Company], 13 February, [online]. http://smartreality.co/five-ways-construction-industry-will-benefit-augmented-reality/ [accessed 13 January 2015].

Kelion, L. (2015) 'Windows 10 to Get "Holographic" Headset and Cortana 2015', BBC News, 21 January, [online]. www.bbc.co.uk/news/technology-30924022 [accessed 22 January 2015].

Kennerley, B. (2013) BIM – Changing Our Industry [Homepage of WSP Group], 2 August, [online]. www.wsgroup.com/en/wsp-group-bim/BIM-home-wsp/what-is-bim/ [accessed 2 August 2013].

MacLeamy, P. (2010) The Future of the Building Industry (5/5): BIM, BAM, BOOM! [Homepage of HoK], 2 February, [online]. www.youtube.com/user/hoknetwork#p/u/38/5IgdcCemevI [accessed 10 September 2011].

NBS (2015) Digital BIM toolkit [Homepage of NBS], 2 January, [online]. www.thenbs.com/bimtoolkit/bimtoolkit.asp [accessed 2 January 2015].

Osbourn, D. (1985) *Introduction to Building*. 1st edn. London: Mitchell's Building Series.

Palsbo, N. and Harty, J. (2013) *Quantitative Materials, Dynamic Quantities Material Libraries*. Sheffield Hallam University, UK: International Congress for Architectural Technology.

Pazlar, T. and Turk, Z. (2008) 'Interoperability in Practice: Geometric Data Exchange Using the IFC Standard', *Electronic Journal of Information Technology in Construction*, 13, pp. 362–380.

Pittard, S. (2012) What is BIM? [Homepage of RICS], 16 March, [online]. fat.glam.ac.uk/media/files/documents/2012-03-16/What_is_BIM_1_.pdf [accessed February 2013].

RIBA (2015) The Future for Architects? - Building Futures Source 2015, 10 April 2015, http://www.buildingfutures.org.uk/projects/building-futures/the-future-for-architectsFoldersLastImported [accessed 9 October 2015].

Rooth, Ø. (2010) 'Norwegian Public Clients and buildingSMART.' University Lecture. Copenhagen: Copenhagen School of Design & Technology.

Selck, P. (1974) Ydeevne – hvorfor, hvordan? (Performance – Why, How?). SBI Anvisning 94, (UDK 69.001.3.004.1).

Stroud Foster, J. (1979) *Structure and Fabric*. Part 1. 4th edn. London: The Anchor Press Ltd.

The Economist (2015) 'Energy: Seize the Day', 17 January, [online]. www.economist.com/news/leaders/21639501-fall-price-oil-and-gas-provides-once-generation-opportunity-fix-bad [accessed 30 January 2015].

Yoders, J. (2014) How is Augmented Reality Being Used in Construction [Homepage of lineshapespace], [online]. http://lineshapespace.com/what-is-augmented-reality/ [accessed 13 January 2015].

6 Game changers
International and national protocols

6.1 BuildingSMART® and interoperability: industry foundation classes, principles and practices

During a recent four-day workshop, 40 Architects, Engineers, Contractors (AEC) professionals and students (roughly split 50/50) assembled to partake in an intensive course to tackle a problematic site, whose role had dramatically changed due to a recent development on a neighbouring site, a real project. There were eight groups of five and the assignment was to analyse and programme a solution for the leftover space. Initially there were presentations of the proposals from the master planners and architects, as well as presentations of the various software programmes and methods to be implemented. The organisers, in close collaboration with the architects of title, had developed a brief of functions and performances, which had to be met in the handling of the project.

This was to be achieved and closely monitored by introducing a BIM Checker, to show and to advise on compliance with client requirements of correctly categorised floor areas and parking. Coupled with this interactive reporting was another form of feedback using simulations and analysis to better inform the decision-making process. These included wind simulation, solar gains, thermal performance and daylight factor amongst others. The adjudication parameters were to be: architecture, energy frame, the environmental impact, the collaborative process, the application of software and co-operation within groups, as well as presentation, argument and validity.

The site was a new urban development in Copenhagen. It marked the border of the existing city with a new town and it was bounded by the refurbished campus buildings of Copenhagen University. Across the road was a new, very dense urban development of 124,000m² by Bjarne Ingels Group (BIG) called 'Batteriet' (or, The Battery). It comprises of a plinth with buildings of up to 18-storeys, including a mosque with minarets, in a cascading architectural interpretation of an Alpine landscape.

The programme required a total area of 9,500m², of which 5,000m² was student residential accommodation, 500m² a kindergarten, 1,000m² parking and 3,000m² commercial floor area. It was to have an energy class for 2020, with a daylight factor of 2 per cent for residential, 7 per cent commercial, 5 per cent kindergarten (with 2 per cent for all ancillary offices), under an expected minimum 1,500 hours of sunshine per annum.

The organisers set up a web-based file hosting service that used cloud computing that allowed users access and sharing of fully synchronised files across the Internet, and a group folder for each group. All information, inspiration, lectures and software links were placed here. The group box was a place to work.

Autodesk, Graphisoft and a few other software producers, with support, were at hand to help install and start up all the programmes we were to use. Some lesser-known products were niche markets and, it must be stressed, not all were used, given the short period we were engaged in, but the biggest players were Revit, Vasari and Ecotect. From these, an IFC model could be uploaded to the Dalux's BIM checker (a server), which in turn generated a report to test compliance. The eventual winners also used Tekla, Robot and Sigma, having also knowledge of Inventor and Rhino.

What was learned from the workshop was that co-operation and collaboration were critical to achieving substantial results. In some of the groups where there was resistance to placing work on the Dropbox, there was a price to pay. While it is easy to see it now, these groups did not have trust, in a professional sense, in place. Largely this was due to architects (both practicing and students) not wishing to relinquish fledgling concepts until they were fleshed out. This was typified by one architect removing/deleting their work after the final presentations from the Dropbox, rather than letting the work be archived for later analysis.

The better performing groups had a majority of technologists (3:2) and while it is true to say that the workshop was very orientated to the implementation of technology, those architects who attended came with an admirable purpose, but resorted to type under the intense pressure and nature of a workshop. This underlines one of the issues endemic in the industry.

Jan Søndergaard, an eminent architect and professor in Denmark, presented a lecture during the workshop where he beautifully distilled the essence out of a project for 'Dacha on the Volga', north of Moscow. It was reminiscent of his summerhouse district and drew heavily on the plot patterns and typologies ingrained there. But significantly, he also stressed the need for a mechanism to exclude other stakeholders from the design until the essential elements were in place. He drew an implicit observation that in excluding a rack of unqualified personnel, he kept control of the layers of the design as they unfolded. He also lamented significantly, in the aftermath discussion that the days were gone when architects 'were best friends [and dined] with the client'.

During the follow-up debate, this exclusion was defended by architects but attacked by those who were excluded. The conclusion was that this is one of the reasons why it is difficult for other stakeholders to work with architects. That increasingly the architect is brought in for a specific purpose, but removed as quickly as it is politely possible from the supply chain afterwards. It was a very dignified debate with openness and honesty from both sides. It also redoubled everyone's efforts to build and mend the bridges in the workshop.

In Anthony Vidler's book on Claude Nicolas Ledoux (Vidler 1990), architectural education in the eighteenth century was broadly described as having few or no standards, and no uniform regulatory codes. The profession, unrestricted by the apprenticeship system of the Middle Ages and Renaissance, had a variety of roles, notionally led by the 'Architecte du Roi', in respect to the general practitioner, and the 'entrepreneur des batiments'. Most of them would speculate and invest, engaging in both design and construction, but membership of the First Class in the *'Academie'* forbade mixing the roles of architect and contractor to preserve the distinction between commerce and the liberal art of architecture.

This small difference could be said to be critical in one of the architectural profession's difficulties today, namely in delivering projects on time and to budget. This separation is also significant of the construction team and the differing roles to be played. The erosion is also compounded by the fact that architects are increasingly finding that they are in fact subcontractors to the prime supplier as client (Worthington 2005).

From the evangelistic viewpoint, this is the clarion call, but from the practical position, there are many other issues. Primarily there is ownership. Who will own the model, who will manage the model, and who will co-ordinate the model's passage through its turbulent growth? In the Gehry case (see Section 3.2) it is a star architect and in such lofty situations those choosing or succeeding to work with him have identified this behaviour as a type of work and have accepted its challenge.

One area where this had to work in the workshop was in meeting the brief requirements. Because of the BIM Checker there had to be teamwork. Once the functions and adjacency diagrams were in place, the rooms had to be 'instance coded' in precise terms for uploading. Once the project's form was made, it had to be tested, to check shading, solar gain and energy consumption.

Then the project's programme had to be approved by the BIM Checker, so that the building's form could be tweaked and manipulated. Accordingly, this meant it was possible now to make architectural statements addressing these issues, backed up by a checking system, and this was empowering. Essentially, what was achieved in four days might pass for a college's semester's project in other circumstances. Granted, it only stretched over the 'Concept' and 'Developed Design' work stages, but the gist of it can be appreciated.

Had it continued into Technical Design and Construction then programmes like Sigma (price book software) and (MS) Project (timeline software) could have delivered more of the 4D and 5D that is happening more overtly in the classroom and increasingly in the industry. With programmes like Navisworks, Tekla and Solibri, the project could be both timeline sequenced, thoroughly checked and fully visualised for all stakeholders.

6.2 Formal schema for classifying and organising data for whole life use

With the modern movement, and new materials like reinforced concrete and steel, the lineage of drawing intent was broken. Increasingly, the know-how that was part of the apprenticeship served by all craftsmen moved from the building site into the drawing office. Here the draughtsmen began defining what, and how they wanted buildings made (Barrett 2011).

Another twist happened with the introduction of CAD into the drawing office after the first oil crisis of the 1970s; a fission occurred between the management and the technical staff where there was no appreciation of how digital draughting was being applied, and the seeds of mistrust were sown. There was also a challenge for senior design team members in implementing new design technologies and adopting new practices (Eastman *et al.* 2008). It took nearly a generation to heal this gulf but arguably, now it is being usurped by BIM (Tse *et al.* 2008).

This can be seen with the various parties working together to produce a building. Previously, light tables would be used to correlate the various tasks and, more recently, overlays of digital drawings could provide a method for formalising the process being undertaken. Both methods involved the checking and cross-checking of other disciplines' work, with the appropriate action being taken to harmonise and synchronise the work.

But the light table does not even feature in the cartoon film industry today and, like the balls of twine that quantity surveyors used to use to take off measurements, has long been consigned to the bin. More common is the emergence of technical meetings, which now occur on site, often in parallel with the architect's site visit, but which may be chaired and run by technologists. These meetings usually involve the subcontractor and the technologist, who puts the work into its context, as well as coordinating sequencing of first/second fixes required to complete the work.

This situation meant that much of the problem solving and rectifications occurred on site, with the ramifications of architects' instructions, change orders and requests for information (AIs, COs and RFIs) ensuring that the project was both late and usually over budget. Indeed, there festered a culture of bidding low, with aggressive litigation to drive costs up and delay handover, which goes beyond the scope of this work.

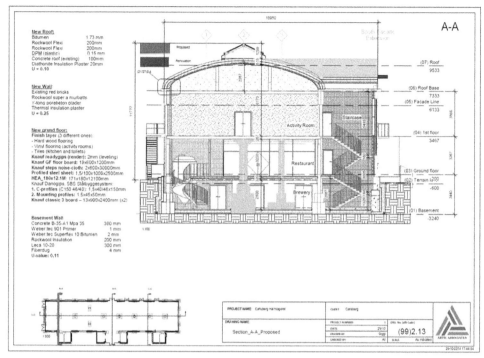

Figure 6.1 Carlsberg Halmlagret project – cross section

As long ago as the 1980s Philip Bennett, writing in *Architectural Practice and Procedure* (Bennett 1981), painted a wholesome picture of the architect in full command at the helm of building procurement. If you wanted a building, you appointed an architect. The architect established the design team, tendered the work and administered the contract.

At that time, most architectural institutional codes of professional practice (RIBA, for example) forbade competition between architects for work. This had the professional ethic of presenting a solid front to the public that upon becoming a member, professional capabilities were guaranteed and certain standards assured through a code of conduct or similar. The market (in the mid-1980s), in challenging this alleged monopoly, forced the introduction of competitive fee tendering for appointments.

While appearing straightforward in its consequences, it also had the effect of polarising practices into specialisation and niche markets. For example, if a firm had several hospital designs under its belt, it had earned a reputation that it could use to its advantage when fighting for the next contract. The image of a pipe-smoking man on a high stool behind a mechanical draughting board, being able to switch from a house extension to a hypermarket master plan, was over.

Ray Cecil, in *Professional Liability* of about the same period, bemoans the carving up of the architect's cake by the 'proliferation of consultants, each of whom has created an institute or association to protect and promote their interests' (Cecil 1989: 130–131). At that time, these consultants included quantity surveyors, various engineers, landscape architects, town planners, and different designers, as well as consultants on sanitation, security, fire, space, acoustics, traffic lobbies, etc., not to mention energy consultants, environmental lobbyists and health and safety officers that we have today. He ends his work shrouded in gloom as to the prospect of the depleted architect's role in the future. Parallel to these developments has been the emergence of project and construction and design managers and technologists (as distinct from technicians), who have also identified a wedge of the architect's rich fruitcake as being fair game.

Various forms of contract, from design and build to partnering, have also reinforced this situation. A traditional contract is becoming rarer, and some would go so far as to say that architects might become marginalised, being hired in a subcontractual role, for a limited design package, and then removed or managed as a perceived risk. So where has this shift come from and where is it going?

Partnering is largely based on trust, dialogue and openness creating a shared culture with trust relationships having common goals, expectations, values, efficiency, innovation and quality. The biggest difference between partnering and IPD is that partnering typically works with short-term alliances, while IPD is for the longer term.

But that aside, large general contractors are increasingly taking the lead role in larger projects, assembling the procurement team and running the contract. Interestingly, too, many multinationals and branded companies are not buying and building their property portfolios, but rather they are commissioning space and leasing it so that they are not responsible for the maintenance and operations for which they otherwise would be liable. This means that there is a de facto inferred life-cycle assessment being implemented, where the onus is back with the large contractor to provide a total enterprise solution. Here, the market rather than legislation is driving the situation (Erkessousi 2010).

In mapping new methods with a view to collaboration the AIA, in its *Integrated Project Delivery (IPD), What, Why and How* (Eckblad *et al.* 2007), drew attention to the fact that in 2007, 46 per cent of architects were unsure about improving current design and construction processes. Furthermore, there was a 27 per cent misalignment between owners (88 per cent) and architects (61 per cent) concerning inflation in construction costs.

Later, the AIA presented a chart showing that approximately two-thirds of architects, engineers and owners had high or good levels of comfort working with each other, but that construction managers dropped to under 50 per cent, a figure lower than was recorded by contractors or sub-contractors. The inference here points to the comfort zone that the design team had nurtured for itself, leaving the construction manager to clean up after them, and be saddled with the blame.

For as long as deliverables remain in a printed-paper format (or even PDF) then the consequence and bottom line is double work. By this, I mean that when design team meetings, contractor site meetings, and all the meetings in between happen with each participant providing paper copy, then this is a recipe for disaster. There are tendencies for open collaboration today, but there is also strong resistance, with copyright and ownership being the main issues, along with payment and limitations to use (Williams 2009).

Traditionally, the architect would sketch and scheme a design and the main disciplines would attend initial design review meetings and by and large reserve space in the grand order of things, so that when the dust settled on the architect's design they could be activated and slot into their respective roles (Shiratuddin and Thabet 2003). This meant that when the architect had set the building out with grids, main arrangements and general form, the other consultants would take the drawing release set and set their own in-house draughting team to prepare a set of drawings reflecting their concerns. In due course, another meeting would be held where all sets would be co-ordinated, and the architect and consultants would check and cross-check each other's work.

Where a building type was well known or common, this strategy had a strong possibility of success, but where some of the roles were new, risks would invariably arise. Also, given that there are only so many plans, sections and elevations that can reasonably be taken of a building project, blind spots had a higher chance of occurrence. But also as the design progressed it was not uncommon for there to be small changes due to building regulations or developments in the brief (by the architect), and the engineers would delay or forego updating each amendment out of a sense of propriety.

Often these sets of drawings could be out of synchronisation for long periods, but this was largely acceptable by the team members, so long as they all remained on board. Projects often went through periods of inactivity due to the client, the authorities or any number of things and, typically, when they became active again, the jagged learning curve raised its head, meaning that the collective memory was not as crisp as would be desirable.

Alternatively, the scope of the building project might be beyond the competence of some of the team members, giving rise to problems on site. Most notoriously, there are clashes and collisions of building elements or components occupying the same space. Less obvious is the sequencing, timelining, or order in which the building elements or components are constructed. The logistics of when things arrive on site and are installed requires careful planning to avoid rework. Lead-in and supply chain ordering can have massive effects on the smooth running of the site if components are not ordered in time or managed regarding their arrival on-site, whether piecemeal or in bulk.

Boeing, the aviation company, is known to have commissioned a virtual robotic figure of human proportions in order to check accessibility for all assemblies in their planes (Gehry 2008). Increasingly, if a building was not rectangular then the risk of cramped spaces or changes in the behaviour of components was not tested. In general, all of these scenarios result from the abstraction of the building project to two-dimensional media. The virtual robotic figure was an attempt to address the geometrical issues but nothing more. The third dimension's impact was largely visualisation and as such treated poorly, and often outsourced as an extra.

The implementation of BIM, it is claimed, reduces the number of hours spent on construction documentation on any project. A skilled designer can realise the intent and detailing required, which results in much less third-party support than previously thought. 'Details, material selection and layouts need only be defined once and can be propagated to all drawings …', where the biggest saving is in billable project hours, for example, an intern architect can go from 320 hours to 96, representing a 233 per cent decrease in demand (Eastman *et al.* 2008: 203).

Reverse engineering this statement, it is arguable that the development and production of a project consumed many hours in cross-referencing work. It is something that is continually being addressed within firms, usually by building teams that comprise people who are good at a particular role or thing, and keeping them locked in that role or process.

6.3 Paradigms for data modelling and management

Synchronisation of data has to become integral to the design process. Uninformed design decisions can no longer be countenanced without consequence. And these consequences encompass the right to practice, as well as the ability to practice if it is not already assumed. 'BIM is a collaboratively generated and maintained, data rich information source for the life of the design process and beyond' (Malleson 2014: 12). Because of the technologies, the amount of inaccuracies and conflicting information can be reduced, specifically in variations, alterations and delays.

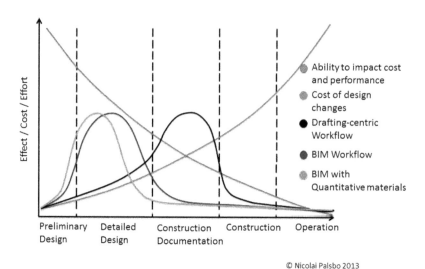

© Nicolai Palsbo 2013

Figure 6.2 Effort against work phases

On adversarial issues within design/construction teams, Patrick MacLeamy went as far as to say:

> in fact today's architects spend about 75% of their time on non-design tasks, practicing what I call defensive architecture. As a result design suffers from lack of attention. Not enough time is put into thoroughly vetting the design to be sure it absolutely suits the client's purposes.
>
> (MacLeamy 2010)

This narrative comes from his infamous curves talk, comparing time over effort through the construction process. Such non-design tasks include regathering and organising data inputs from other disciplines. MacLeamy claims that if the fragmentation of the building industry is to be dealt with, it must be by removing the adversarial culture played out by the different stakeholders in a project (Tranum Mortensen 2014).

Another important aspect is the introduction of soft landings, and to an extent the new (RIBA) work stages, 'Strategic Definition' and 'In Use', which go beyond the mere procurement and make for a longer engagement for all stakeholders (Harty 2014). The proper use of soft landings, when applied to projects, means that operational outcomes are understood so as to better match design intentions, which are defined in the strategic brief. They bring the client much more into focus and offer a method of continued contact. They open up aftercare activities to observe and fine-tune handed-over facilities. This creates shared risk and responsibility, which can only help build trust and collaboration across the construction spectrum. Of course, an extended contract increases fee potential.

The core objective of the 'Strategic Brief' phase is to identify the client's business case, so that sustainable decisions are made. The core objective of the 'In Use' phase is to undertake in-use services so that the facility is used how it is intended to be used and so that EBD can begin to have an impact (Mortensen 2014). EBD is the realisation that design options should be based on researched evidence rather than on the intuition of the designer (Ulrich 2001).

This fuller engagement compels all stakeholders to consider and formulate their input for the longer benefit of the facilities' life cycle. Better simulation, analysis and performance then enter the fray. Buildability improves: both sustainability and digitalisation become the driving force for better communication among the customer (both client and user), design team, contractor, subcontractor and supplier (Manthorpe 2014). Bringing consultants, developers and users closer together integrates design better (Gravad and Brenøe 2014).

Traditionally, contracts defined an agreement between two parties, in isolation from each other and with scant regard for collaboration. The contracts discouraged the sharing of models and data, amid concerns for design shortfalls, and preserved liability, with no regard for intellectual property rights. Since BIM encourages fluidity in the design process, other issues regarding distribution to third parties, as well as ownership, are drawn into the mix. Furthermore, data is easier to move and reuse in other projects.

Lip service has often been paid to collaboration, and readily, too (Smith and Tardif 2009). There have also been very successful procurement teams with very successful building projects. But there has never been a repeatable mechanism to propagate and develop these relationships. The industry is noted for its fragmentation and even if the same disciplines or practices work together on subsequent or continued work, often the personnel can be switched out for whatever reasons meaning that the learning curve is jagged with drops in the acquired knowledge accumulated over the course of the project.

Furthermore, when requested to collaborate by the client (or out of good common sense) often there is no device to allow this to happen, meaning the adversarial nature of traditional contracts and appointments hinder such co-operation. Sometimes it is seen as correct procedure to nay-say and remain aloof from the proceedings, divorcing the commerce and the liberal art of building as mentioned earlier.

Cicmil and Marshall (2005) elaborate and clearly elucidate a scenario of pseudo collaboration, where a two-stage tender is hopelessly inadequate due to the intransigence of the Quantity Surveyor (QS) in their perceived role of adviser to the client, rather than a deliverer of the project. There is no mechanism in place to allow the QS to enter into a collaborative state with the main contractor, and no desire to either. In an earlier study, Cartlidge (2002) probably summed it up best in a condemnation of the then relationship of QSs with clients, recommending 'quantity surveyors must get inside the head of their clients' (p. 33).

In providing a legal basis, contracts only establish a consensus in the allocation of risk and liability. In a collaborative context, BIM is intended to provide benefits to all parties, rather than loopholes for damages by negligent errors. The emergence of Levels of Detail/ Development (LODs) and Levels of Information (LOIs) goes some way to alleviating these concerns.

What these mechanisms bring to the project team is the ability to monitor and act with precision, in the flux of the process. They offer a protocol where all stakeholders can partake in full confidence of the design's status, without acting out of turn or negligently. Liability and permitted purpose make actions more robust while ring-fencing responsibilities in a fragmented market. As Simon Lewis says:

- it prepares managing the processes and procedures for information exchange;
- it initiates and implements the project information plan and asset information plan;
- it assists in the preparation of project outputs, such as data drops; and
- it implements the BIM protocol and updates the Model Production and Delivery Table (MPDT) as and when necessary.

(Lewis 2015)

Essentially, the LODs map the progression of the building elements from generic forms to as-built assets, while the LOIs track the documentation trail from initial scoping statements to fully specified texts. This process builds into a digital toolkit, which will transform the procurement of both buildings and infrastructure. The toolkit will make available a digital plan of work, and a classification system which incorporates definitions for construction objects at each of the delivery stages throughout the life of a built environment asset (NBS 2015).

6.4 Procurement methodologies, BIM and the EU

To reduce the cross-referencing in practice several issues need to be resolved, some are to use inducements and offer professional services compensation, to offer project delivery performance and/or business enterprise performance optimisation incentives. By doing this, mutual respect and trust can be nurtured and benefits and reward can be encouraged, while more collaboration comes into the innovations and decision making.

Furthermore, it is not unheard of for firms to amalgamate two or more disciplines in an effort to gain an advantage in bidding for and delivering work. This could be architects and quantity surveyors or the many types of consulting engineers that we find, where the purpose is to remove potential barriers or internal conflicts to the procurement process (Smyth and Pryke 2006, 2008). With new methods of tendering and partnering different consortia often find it advantageous to come together on a project-by-project basis. Prequalification means building up a track record in predefined core competences. There are many strategies at play, and many responses. Some consulting engineering firms are expanding to offer one-stop-shop solutions, so that the client only has one avenue of redress.

In the USA, in the Association of General Contractors' (AGC) own guide to BIM implementation, they blast the first two myths: that BIM is only for large projects and large contracts (Ernstrom *et al.* 2010). They identify the benefits in a no-nonsense style, mentioning collisions detection, visual communication, fewer errors, higher reliability, better 'what-if' scenarios, better end-product for clients, and users with fewer call-backs (snags), meaning lower warranty costs.

A reimbursement contract is usually used in refurbishment work (where it might be difficult to assess the cost of work), in which case it can be used to reimburse the contractor for the costs, plus a fee to cover overheads and profit. Digitalisation here, beyond the previous method, would be a checks-and-balances means to justify the costs. But laser scanning is also beginning to appear in this niche as a means of control, either at commencement of the works to record the existing context, or more dynamically as a regular or daily method to track the progress of the work against the virtual model. This is very advanced but, with the development of GIS technologies, will push the augmented reality aspect of things.

Looking at RIBA's work stages and Plan of Work (RIBA 2013), many combinations of procurement are mentioned, including: fully designed project with single-stage tender or with design by contractor or specialist; design and build with single- or two-stage tender; partnering contracts; management contracts; and public-private partnerships and private finance initiatives. Within these options there are appointments and selections, and input and output packages with requirements and proposals. In a recent survey, 38 per cent of architects and only 22 per cent of contractors found design-bid-build to be the best system to reduce costs. It rose to 43–50 per cent with design and build. Interestingly, 'Construction Management at Risk' owners found that 33 per cent of their projects were under budget with 60 per cent being satisfied with their project's progress (Bernstein and Laquidara-Carr 2014).

Depending on the size and complexity of the project there are a number of ways that its design and construction can be undertaken (Müller 1997). First, there is the traditional contract, using a standard form. It usually requires the contractor to carry out the construction according to the drawings and specification drawn up by the design team, where the work is supervised by the architect. Digitalisation in this context is a simple translation to the new media with little or no new input. All parties maintain their independence and retain their core competences.

With a fixed price contract the contractor agrees to construct the building as specified in the drawings with an unpriced bill of quantities for an agreed sum, by an agreed date. The contract allows the contractor to claim additional costs for any variations to the specification. The allowance can also be claimed for an extension of time for delays beyond the contractor's control. It can be for the whole contract or a section of work, or it can be applied to a unit rate, where the price is fixed but the amount of work is unknown.

This is an area where contractors are fast becoming the drivers of the digitalisation process, with the AGC in America recommending to their members not to bid on non-BIM'ed work or, in the worst case, to build their own model before bidding in order to have better control on estimates and processes (Young *et al.* 2008). This is a significant paradigm and a move that is changing the drivers of BIM adoption.

In design and build contracts, the contractor is responsible for the design, specification and construction. It may be on a fixed price or cost reimbursement basis, which is either negotiated, or subject to tender. It is normally used for standard or repetitive building types, where the contractor has previous experience, resulting in savings for the client. This is very appealing to digitalisation, especially where there is duplication of the building type with serial clients. The benefits and return on investment make this very attractive for all involved.

Develop and construct contracts are similar where a design team is appointed to produce concept drawings prior to going out to tender. The advantage is that the developer is only dealing with one source that has sole responsibility for the project's design and construction. This means that there is an awareness of the financial commitment prior to the commencement of construction.

In management contracting, the design team specifies the building requirements, and specialist subcontractors are supervised and co-ordinated by the management contractor to carry out the construction. For this, the management contractor receives a fee, which may be fixed or a percentage of the contract cost. It is generally used on complex projects that require a short contract period, which must have flexibility for modifications during construction. Digitalisation has a huge benefit in this fast track method but requires a highly motivated personnel and digitally competent team.

Subcontractors can enter into contracts with the management contractor to carry out specific work, where it is the management contractor who has a contractual relationship, not the client. If any problems arise, it is the management contractor who must pursue for a remedy. Subcontractors are normally appointed by competitive tendering based on the drawings and bills of quantities. For the contractor the biggest advantage here is that there are very few risks, as they are guaranteed a return of costs and they do not have the problems associated with the employment of labour. It is important for the project manager and quantity surveyor to control costs, since the management contractor has no incentive to control costs within the cost budget, although incentives can be introduced. For the developer, work can begin as soon as the first few work packages are produced, so allowing design and construction to overlap.

The interesting part here is where the model is made available to the subcontractor, meaning the contractor has been given the model. This is rare at the time of writing, but in isolated cases and pilot studies, subcontractors were at first most reluctant to engage a model, citing all kinds of excuses ranging from beyond their scope to tried and trusted traditional methods. But having complied, there was a watershed moment of 'how had they not been doing this sooner'. It becomes particularly relevant with regard to augmented reality.

Construction management is similar in most respects to management contracting, except that the contracts are made with the client. The construction manager is employed to manage the construction work. This system is used mostly on large, specialist technical projects. The payback here is that the on-site phase can be carefully monitored and fine-tuned in the model, meaning that the model is as near to the built reality as possible, and is ready for handover to the facilities managers upon completion. The real bonus here is project certainty, both in terms of time and budget, and the continued relevance of the model under operations and maintenance (life cycle).

Project management is also normally used on larger developments. It is becoming increasingly popular but, more importantly, it is reducing the architect's influence within the construction industry. The project manager can be an organisation or an individual, who guides the client through the procurement system, appointing the construction team and controlling the project. Usually appointed on a fee basis, the project manager is not dependent on the cost of the contract. This tends to ensure that the project manager works solely in the client's interest, as there is no commission. The potential here is that the divorcing of the architect and control of the job leaves the door open for a manager of sorts.

Whether the role is filled by someone from the construction industry or by someone who comes into the industry with pure management skills is up for debate but interestingly with regard to the building information model for which there is a need for the management of sharing, integrating, tracking and maintaining data sets, this offers the opportunity of a new and awesome task.

Partnering covers both public-private partnerships (PPPs) and private finance initiatives (PFIs), which are special relationships between contracting parties in the design/construction industry (Erkessousi 2010). They positively encourage changes to traditional, adversarial relationships to more co-operative, team-based approaches. By this, they promote the achievement of mutually beneficial goals while also preventing major disputes.

PFI is an arrangement where public sector assets and services are acquired through private sector funding, thus reducing government/public sector borrowing. It is a procedure where the public sector sponsors or establishes a business case strategy. In Europe, the project is then advertised in the *Official Journal of the European Union* (OJEU). Through a prequalification process the bidders are then short-listed, and a consortium is usually set up specifically for the project, forming a Special Purpose Company (SPC). The contract can then be awarded.

The SPC is usually financed by 10 per cent from the company, with the remainder coming from the financial institutions. This is then recouped over the next 25–30 years from tolls or service charges upon completion. There is good potential for high returns, and it gives continuity of work and offers involvement in the design phase. It means that it is highly buildable, because of the make-up of the stakeholders, and it offers more control over the programme than might be possible under traditional methods. On the downside is the initial bidding costs process, which can also be long. It is a very competitive market, tying up many resources initially, and is also quite complex and demanding. Contract terms can also be very onerous and penalising, and it usually comes at a fixed price for the contractor. Their main features require firm commitment from top management, encouraging continual improvement while also allowing time for the benefits to emerge. The basis for these mechanisms is to break the traditional mould, which engulfs the industry, and to provide a forum where new talent can showcase their worth.

They are also based on the equality of all partners, with an interest in mutual profitability. In total contrast to the traditional contract, partners have free and open exchanges of information. This lends itself well to BIM and opens a bigger scenario, called IPD.

Not so obvious is that IDP tries to keep project teams together, which is done with well-thought-out incentive schemes. The top reason for this is that it reduces the learning curves that are otherwise necessary, while eliminating the sawtooth information drops through work phase handovers or similar. Great value is therefore placed on long-term relationships, and an environment for long-term profitability exists where overall performance can be improved.

With all parties seeking a win–win solution, everyone understands that no one benefits from exploiting each other. Crucial to this process is the problem-resolving ethos that it engenders. By having mutual objectives, the door is open to consideration, as a concept, of each other's worth. Ironically, this was missing previously. Trust and openness is encouraged to openly address problems, so that innovation is embraced positively. Each partner becomes aware of the others' needs, concerns and objectives and is interested in helping to address them.

Prime in its objectives are improved efficiency and a reduction in costs. Dependable production quality then leads to speedier construction and more certain completion time. Longer-term benefits include better continuity of workload and a more reliable flow of design information.

The shared risk has both positive and negative connotations. In short, the team works much better together, but in contrast, one bad apple can turn the whole barrel. With the reduction in litigation, there is the knock-on effect of lower legal costs and exposure. There is also a lessening or removal of large contingency sums, with better decision and problem-solving systems. This, in turn, can equate to savings of approximately 5 per cent on project costs, 6 per cent for clients with profits of up to 9 per cent accruing according to Constructionsite (Constructionsite 2010).

Where problems do occur, these derive largely from the fragmented nature of the construction industry, typically from the low-bid mentality or from corruption. At the other end of the scale are issues of intellectual property and complacency (Williams 2009).

Once an IDP process is primed and under way, there can be earlier involvement of key participants and earlier goal definitions, enhanced by early planning, more open communication opening the way for appropriate technologies and better informed leadership. In a survey (conducted by the AIA in 2008), respondents were asked to prioritise issues seen as barriers to adoption of IPD. The 'experienced' in IPD methodologies, compared to those categorised simply as 'knowledgeable', expressed less concern about the barriers to adoption. These findings suggests that the 'actual risks' associated with IPD are less significant than the 'perceived risks' expressed by those who are not fully committed.

Finally, George Miller argues that to reduce these barriers, one should consider a facilitator or an outside catalyst. If it is difficult to break away from the historically adversarial behaviours typically found in the design and construction industry, then such an adjunct consultant might smooth the transition over. Such a facilitator might be of great use to teams, trying to ease away traditional boundaries and enhance collaboration.

Once your own house is in order, the next step is to find like-minded neighbours. Much the same process that was described previously applies equally here, and even more so because between practices it is more difficult to enforce or encourage adherence or coherence. The risk factor is considerably higher because all parties are working at risk, meaning ultimately they inextricably depend on each other. If the whole portfolio of a firm was so exposed, it could jeopardise the firm's existence, in a weakest link scenario.

Traditionally, tendering for work, or prequalification, meant setting manpower and resources against the job in hand. In splendid isolation, and with disregard to the other teams, a conservative estimate was usually put forward. This meant that if changes occurred during procurement, with a model being initiated and passed around, then clearly the same manning was not required and the project was consequently overmanned. Rather than reducing the commitment pro rata, often there was an attempt to bully or bluff the client into accepting the agreed manning levels and paying out on the pre-planned mobilisation. Such a situation was usually sold with guarantees of compliance and cohesion, and the extra manning was then set to checking and cross-checking as per usual.

This is a self-perpetuating vicious circle which encourages the project architect or engineer to repeat the exercise next time there is a bidding process. It provides work for the firm and meets in-house manning targets and expectations. It will only be when the client demands a BIM model and collaboration that these tendencies will die out.

There are many ways of reducing human errors and prime among these is experience. But expecting clients to foot the bill for gaining experience is becoming a contentious topic (Barrett 2011). Actuaries mean that it would take 15 years to fully train an architect with the resulting uproar of this being too long and unpalatable. Compromise is important as there will always be continuing practice development and so forth, but another avenue is available, namely employing the same model. This reduces the double work, which is critical to addressing the shortcomings mentioned earlier, but a new mindset is required to implement it.

Collaboration not only reduces errors, but it also improves communication and understanding of the other roles in the procurement process (Ribeiro *et al.* 2010). Traditionally, this was achieved through light tables with tracing paper and overlays in CAD digital formats. Again, only the material presented could be checked, meaning a more meaningful method was required. This is also reminiscent of the sawtooth knowledge drops incurred through a project when there are several handovers of information (at design, procure, build and manage) (Eckblad *et al.* 2007).

The classic situation on a building site is the services person installing an air duct on the third floor that needs to go through a structural concrete beam. The services person calls the site hut and the site architect comes up. The site architect goes back to the site hut and resurrects the drawings for architect, engineer and services to see if this is indeed the case and if there is something to be answered. If there is a fault or blame to be apportioned, the necessary paperwork is set in train. If it is an oversight, all relevant parties are summoned either to an ad hoc meeting or it goes on the agenda of the next site meeting.

Either way the situation has to be resolved, which means new work, a lot of paperwork and added costs and time delays. By the time a meeting can be arranged with all parties concerned a week can go by and by the time the blame is apportioned and the costs are distributed thousands of pounds can be added to the job. Something has got to give.

Requests For Information (RFIs), Change Orders (COs) and Architect's Instructions (AIs) are mechanisms to seek/issue further drawings, details and instructions beyond the documentation included in the contract to cover unforeseen things during the building phase. They are conditions laid down to typically correct discrepancies in documents; comply with statutory requirements; correct setting out work; make good work, materials or goods not in accordance with the contract; exclude people from the site; and so on (Bagnall *et al.* 1990).

The obvious antidote here would be to have better and more complete documentation before going on site. To resolve these issues, a better way of finding and correcting mistakes is required. This takes us back to the idea of minimalising double work and reducing human error. In a case study, all the site excavations, foundations and basement work resulted in a single AI being issued for one borehole that was missed. This was achieved with better tools and better collaboration earlier in the procurement phase. All parties have moved to better co-ordination of work, with standardised methods of measurement and common classification systems, giving better schedules and specifications. But sharing the data, as is now possible with BIM, is not happening as quickly as it should, given the tacit improvements it provides, and this raises many concerns.

In a Danish project, six houses were sponsored by Realdania to address this CO_2 situation and to take the first steps to open the debate about how we best deal with it (Realdania 2014). The first house of the series was the 'Upcycled House' where it was the intention to upcycle waste materials to new or better products. Upcycling can be described as a process that converts waste or useless materials into new products or materials of higher value.

Anders Lendager (the architect) initially thought that the target of 65 per cent CO_2 that could be saved was unrealistically high, but when the whole project was completed, it turned out that it had an 86 per cent better CO_2 balance when compared against a benchmarked house. With that in mind, it was questioned why no one else was working with upcycling. Why is it not part of what all architects engage in, and why is it not included in the Building Regulations, that a certain percentage of building materials should be upcycled?

The second house was the 'Low Maintenance House' (using 'Traditional Methods'), where essentially the life-cycle period was extended to make material choices more attractive. Both the expected service life and the overall maintenance are key factors when it comes to reducing a house's CO_2 emissions. Although the energy consumption for the operation of a house is still a dominant factor in the overall CO_2 balance, it can be said to be well catered for within the building code requirements for building energy-efficient increases.

This means that the focus should now be geared more towards the embodied CO_2 in the house and its materials, because they will account for a larger percentage of the house's total

Figure 6.3 Upcycle House

Figure 6.4 Low Maintenance Traditional House

CO_2 emissions. Therefore, it makes good sense to increase a house's lifespan from the typical life of 50 years to three times that. If a house could achieve a lifespan of 150 years, without requiring maintenance in the first 50 years, then savings of CO_2 can be made which would equate to building up to three new houses (Leth and Gori 2014).

The second 'Low Maintenance House' (using 'Technological Methods') tackles the issue using innovation and the latest technologies to build their argument. Here a double facade is employed in a modern version of a typical Danish house. It becomes a glass shell that grows around the house as a building envelope, providing effective protection of the house's organic and biodegradable materials:

> It makes very good sense to reduce the palette of materials used. Today we increas-ingly rely on technological solutions and we have a lot of technological equipment in our houses. But equipment constantly undergoes continuing development meeting ever stricter requirements, and this can mean that a technology may become obsolete before its time. Our strategy has been to reduce the number of gadgets as much as possible, by using long-life components and integrating technological installations, so that they can be easily removed or replaced.
>
> (Arkitema Architects 2014)

Most homeowners implement ongoing changes to their house over time, as the family grows (or even becomes smaller). There might also be a need to just account for changes in tastes. Whatever, it requires time, resources and CO_2 to customise a house to these changing needs. Refurbishment, alterations and extensions have an intrinsic cost in CO_2 terms, especially when they have not been anticipated. The 'Adaptable House' demonstrates a house born to be

Figure 6.5 Low Maintenance Technological House

flexible to changes throughout its life, which then reduces material consumption and therefore the amount of CO_2 emissions associated with change.

Each and every time materials are replaced, CO_2 is discharged. To address this, the Adaptable House can be easily altered to minimise the effect. Its construction emits 10 per cent less CO_2 than the benchmarked house (4.5 compared to 5kg CO_2/m2/pa). This means that the savings from typical changes of 50 per cent do not cost extra in terms of CO_2 in the building materials (Henning Larsen Architects 2014).

By contrast, the 'Quota House' challenges our cultural mindset by penalising misuse and poor practices. Is it fashionable to live in such a house, would you tell your friends? Where is the benefit, both at the micro level and the macro level? How can such a way of living be introduced and rolled out as the de facto lifestyle?

Well, the physical house has been built and, at the time of writing, is still unoccupied, awaiting tenure. The family that moves in will need to be braced for and forewarned about how to live here, because energy will be metered a certain amount each month. Spend your quota too soon and it will be a bitter end to the billing period. But manage your quota, or even save some of your quota, and a different picture emerges. Now you have a commodity to sell, save or dispose of as you see fit.

Management is key here: proper metering and careful supervision leads to a better lifestyle. Knock-on effects mean that the mindset permeates all, bringing new mores and standards to day-to-day toil. Being positively accepted is critical here: people need to voluntarily buy into its karma and discipline. The rich person needs to be seen as the one who uses least energy.

For this to happen, technologies, whether active or passive, must be simple to operate and based on the lowest common denominator. Of necessity, the interface between the user and

Figure 6.6 Adaptable/Flexible House

the technological system must be uncomplicated. If at all possible, systems should be intuitive to use and the output from the inventory system easy to understand. Ultimately, this maximises efficiency and, most importantly, raises energy literacy (Williams 2012). In the case of the 'Quota House', a wall-mounted monitor has an image of a polar bear on an ice cap. As the billing period grows, the ice reduces proportionally to the occupants' energy use, giving them an idea of their energy consumption and an image to help them stay positive within what they are involved.

The advantages that accrue from this are environmental, economic, social, ethical, and reciprocal (in that they give feedback). In the bigger picture, making technologies simple to operate means they are easy to replicate. Appropriateness, simplicity and even diversity demonstrate these relative advantages. It leads to a symbiosis, which in turn develops resilience in making the delivery of the house possible. Long-term, this is vital to the emergence of the house, encouraging innovation and defining cultural trends, so that it becomes mainstream and acceptable.

Finally, the 'Best Practice House' is the last house of this project, incorporating all the best practices learnt from building the other houses, while building public awareness and hopefully making a statement of intent.

The catalogue house company Benée Houses won the contract to build this MiniCO$_2$ house type. It will provide the lowest achievable CO$_2$ footprint of a standard house. Here, the resulting CO$_2$ reductions will be carefully evaluated against effectiveness and price, seen from an holistic viewpoint over the construction, operation and maintenance. It will be the optimal CO$_2$ reduced standard house, having the same economy that is standard and expected for other houses today. The house is now under construction, and nearing completion.

Figure 6.7 Carbon Quota House

Figure 6.8 Best Practice House

For these kinds of projects to become mainstream, the process and results need to be recorded and published. This is a method known as EBD, and it allows designers to become better acquainted with the size and scope of the problem. It also allows them to develop new expertise and expand their portfolios into new niche areas.

According to a journal from the AIA, there are four levels of Evidence Based Design (EBD) Practice (Hamilton 2003), but architects are usually not taught research methods. For this reason they find it difficult to use, and even more so to perform, research, which is increasingly demanded due to the development of the specialised roles of the architect. The four levels of EBD is a model to differentiate the practice of EBD into four types. Each type increases the level of commitment to be done, and to be used in research on behalf of the client:

- Practitioners interpret the evolving research related to the project and make judgements about the best design in specific circumstances, for example, they use benchmark reviews of other projects and research publications. Most architects work at Level 1, delivering an improved design.
- These practitioners take the next step up the ladder and start to theorise on the outcome of their design solutions and to measure the effects of these.
- Practitioners at Level 3 follow literature, hypothesise on the outcome and measure the outcome as Level 2 practitioners do, but also publish their results and thereby share their information beyond company and client level.
- Practitioners at Level 4 do the same as Level 3, but go further by publishing the results in peer-reviewed journals or collaborating with university professors for highest level of review.

(Mortensen 2014)

In creating a market for near zero carbon homes, several issues must be dealt with so that there is no objection to their adoption. Near zero carbon homes have a need for some sort of parity against standard homes in terms of convenience, life satisfaction, economical costs and, not least, social prestige (Williams 2012). Having a quota must not be seen as a punitive punishment. Rather, the benefits and possibilities of not filling your quota must be stressed, encouraged and rewarded handsomely. The ensuing payoff should be better marketed and carbon must be seen as a currency (needing a bank) with better appeal on the stock market. This harks back to the point that carbon needs proper pricing. BIM can play a major role here in establishing the amount of CO_2 in a project, so that qualified decisions can be made and acted upon.

The market should be allowed to find its own level rather than being politically held back, as when the price of a permit to emit a tonne of carbon dioxide was allowed to fall 40 per cent to €2.81 in 2013 (Carrington 2013). This was so that energy companies, who from 1 January 2014 have to buy credits to offset their poor behaviour, are no longer treated favourably. Prices above €20 are needed to give utilities the incentive to make serious switches to lower carbon energy generation. Connie Hedegaard has vowed to fight on to save the emissions trading scheme (Harvey 2013). There is a Social Cost of Carbon (SCC) and there are Marginal Abatement Costs (MACs), which only become interesting when the carbon price rises (Prentice 2012). BIM, through CO_2 scheduling, can accelerate this process.

In Germany, capital and operational subsidies (feed-in tariffs) have had a beneficial effect. They play a crucial role in bringing confidence to the implementation of greener solutions. Guaranteeing financial returns over 20 years encourages investors and offers much-needed security (Williams 2012). In California, grants are offered to solar home initiatives. Shortened planning processes are offered in San Diego, and Scandinavia generally has implemented change through building regulations.

To reduce emissions of greenhouse gases (GHG), a carbon tax would be more effective than subsidies for windmills and nuclear plants (The Economist 2015). But consumer-led policies are not enough: producer-led policies are needed to increase supply, and this comes through innovation. Innovative technologies should be heralded and promoted, whether by political will or individual consumerism. By implementing code-checking through an IFC model, this process can be more transparent (Rooth 2010).

The backdrop for this is the materiality of the house, the technology and finally cultural betterment which needs to be seen and enjoyed in terms of energy literacy. Society needs to be courted and flattered into adopting these measures so that peer pressure makes a positive contribution. Incentives (feed-in tariffs) seem to be one way of making quotas digestible.

Oil, coal and gas accounted for 32.9 per cent of energy demand, 30.1 per cent and 23.7 per cent respectively (86.7 per cent total) in the USA in 2013, of which 59 per cent of energy was wasted. Between 2000 and 2012, China's consumption of oil more than doubled (The Economist 2015). Amory Lovins, in the same article, suggests that proper building designs and energy storage can eliminate the need for heating and cooling in most climates. Better insulation and energy efficient glazing in his own house in Colorado (where the temperature can reach as low as $-44°C$) has more than repaid itself in eliminating heating bills.

References

Arkitema Architects (2014) Hus 2 og 3: De Vedligeholdelsesfri Huse [online]. www.realdaniabyg.dk/projekter/minico2-husene/hus-2-og-3-de-vedligeholdelsesfri-huse [accessed 28 May 2014].

Bagnall, B., Brett-Jones, T., Dallas, H., Oakes, J., Oakes, R., Pickard, Q., Poole, G., Quaife, G., Townsend, J., Willcock, J. and Williams, J. (1990) Contract Administration for the Building Team, The Aqua Series.

Barrett, N. (2011) The Rise of a Profession within a Profession: The Development of the Architectural Technology Discipline within the Profession of Architecture. OpenAIR@RGU, [online]. http://openair.rgu.ac.uk [accessed 13 August 2015].

Bennett, P.H.P. (1981) *Architectural Practice and Procedure* (Mitchell's). 1st edn. London: Batsford Academic and Educational Ltd.

Bernstein, H.M. and Laquidara-Carr, D. (2014) *Project Delivery Systems SmartMarket Report (2014) – SmartMarket Reports – Market Trends*. USA: McGraw Hill Construction.

Carrington, D. (2013) 'EU Carbon Price Crashes to Record Low' | Environment | theguardian.com, 24 January, [online]. www.theguardian.com/environment/2013/jan/24/eu-carbon-price-crash-record-low [accessed 31 May 2014].

Cartlidge, D. (2002) *New Aspects of Quantity Surveying Practice*. Oxford: Butterworth Heinemann.

Cecil, R.J. (1989) *Professional Liability*. 3rd edn. London: Legal Studies & Services (Publishing) Ltd.

Cicmil, S. and Marshall, D. (2005) 'Insights into Collaboration at the Project Level: Complexity, Social Interaction and Procurement Mechanisms', *Building Research and Information*, 33(6, November/December), pp. 523–535.

Constructionsite (2010) Contract Procurement Methods [Homepage of Constructionsite], [online]. www.constructionsite.org.uk/repository/resource/view_resource.php?id=10 [accessed 10 October 2010].

Eastman, C., Teicholz, P., Sacks, R. and Liston, K. (2008) *BIM Handbook: A Guide to Building Information Modeling: For Owners, Managers, Designers, Engineers, and Contractors*. USA: John Wiley & Sons.

Eckblad, S., Rubel, Z. and Bedrick, J. (2007) *Integrated Project Delivery: What, Why and How*, 2 May 2007, AIA.

Erkessousi, N.E. (2010) How Integrated Project Delivery is an Advantage to the Danish Building Industry, and How it Can Be Executed. Bachelor of Architectural Technology & Construction Management. Copenhagen: Copenhagen School of Design & Technology.

Ernstrom, B., Hanson, D., Hill, D., Clark, J.J., Holder, M.K., Turner, D.N., Sundt, D.R., Barton, L.S.I. and Barton, T.W. (2010) *The Contractors' Guide to BIM*. USA: Association of General Contractors (AGC).

Gehry, F. (2008) Digital Project – Frank Gehry, 26 November 2008.

Gravad, J. and Brenøe, D. (2014) (Inspirational Talk) New Business Methods and Globalisation. Copenhagen: Head of Department, Building, Renovation & Refurbishment. Ramboll Group A/S.

Hamilton, D.K. (2003) 'The Four Levels of Evidence-based Practice', *Healthcare Design*, 3(4), pp. 18–26.

Harty, J. (2014) The RIBA Plan of Work: The Way Forward through Soft Landings, T. Koundar, ed. ICAT Aberdeen 2014, 7 November 2014, RGU.

Harvey, F. (2013) 'Europe's Climate Chief Vows to Fight on to Save Emissions Trading Scheme' | Environment | theguardian.com, 17 April, [online]. www.theguardian.com/environment/2013/apr/17/europe-climate-chief-vow-save-emissions-trading [accessed 31 May 2014].

Henning Larsen Architects (2014) Hus 4: Det Foranderlige Hus, [online]. www.realdaniabyg.dk/projekter/minico2-husene/hus-4-det-foranderlige-hus [accessed 28 May 2014].

Leth, U. and Gori, K. (2014) Hus 2 og 3: De Vedligeholdelsesfri Huse, [online]. www.realdaniabyg.dk/projekter/minico2-husene/hus-2-og-3-de-vedligeholdelsesfri-huse [accessed 28 May 2014].

Lewis, S. (2015) 'The CIC BIM Protocol: Some First Thoughts' – Building Information Modelling (BIM) article from NBS [Homepage of NBS], 2 January, [online]. www.thenbs.com/topics/bim/articles/cicbimprotocolsomefirstthoughts.asp [accessed 2 January 2015].

MacLeamy, P. (2010) The Future of the Building Industry (3/5): The Effort Curve [Homepage of HoK], February, [online]. www.youtube.com/watch?v=9bUlBYc_Gl4 [accessed 10 September 2011].

Malleson, A. (2014) BIM Survey: Summary of Findings. *NBS National BIM Report*.

Manthorpe, M. (2014) (Inspirational Talk) The Construction Industry's Future and Trends, Ethics and the Turnaround from the Economic Crisis. Workshop Presentation. Copenhagen: NCC.

Mortensen, P. (2014) Evidence Based Design in Danish Healthcare. Dissertation. Copenhagen: KEA.

Müller, E. (1997) *The Constructing Architect's Manual*. Denmark: Horsens Polytechnic.

NBS (2015) Digital BIM Toolkit [Homepage of NBS], 2 January, [online]. www.thenbs.com/bimtoolkit/bimtoolkit.asp [accessed 2 January 2015].

Prentice, D. (2012) *Carbon Management*. Course Material. Coventry: European Energy Centre, Centro Studi Galelleo.

Realdania (2014) MiniCO$_2$ Houses – Experimental Newbuild Construction in Nyborg, Denmark [Homepage of Realdania], 28 August, [online]. www.realdaniabyg.dk/projekter/minico2-husene [accessed 31 March 2014].

RIBA (2013) *Plan of Work*. UK: RIBA.

Ribeiro, F.L., Leitão, V. and Ferreira, T. (2010) 'Using Knowledge to Improve Preparation of Construction Projects', *Business Process Management Journal*, 16(3), pp. 361–376.

Rooth, Ø. (2010) *Public Clients and BuildingSMART: Key Issues in Building Policy*. Keynote address. Copenhagen: Copenhagen School of Design & Technology.

Shiratuddin, M.F. and Thabet, W. (2003) *Implementation Issues of Design Review System Using Virtual Environment*. Nashville, TN: ASCE Construction Research Council.

Smith, D.K. and Tardif, M. (2009) *Building Information Modeling: A Strategic Implementation Guide for Architects, Engineers, Constructors and Real Estate Asset Managers*. Hoboken, NJ: John Wiley & Sons Inc.

Smyth, H. and Pryke, S. (2006) *The Management of Complex Projects: A Relationship Approach*. Oxford: Blackwell.

Smyth, H. and Pryke, S. (2008) *Collaborative Relationships in Construction: Developing Frameworks and Networks*. London: Wiley Blackwell.

The Economist (2015) 'Energy and Technology: Let There Be Light', 17 January, [online]. www.economist.com/news/special-report/21639014-thanks-better-technology-and-improved-efficiency-energy-becoming-cleaner-and-more [accessed 30 January 2015].

Tranum Mortensen, O. (2014) Implementation of Building Information Modelling and the Impact on Design Culture. Bachelor Thesis. Copenhagen: KEA.

Tse, T.K., Wong, K.A. and Wong, K.F. (2008) 'The Utilisation of Building Information Models in nD Modelling: A Study of Data Interfacing and Adoption Barriers', [online]. www.itcon.org/cgi-bin/works/Show?2005_8 [accessed 27 November 2008].

Ulrich, R.S. (2001) Effects of Healthcare Environmental Design on Medical Outcomes, Design and Health: Proceedings of the Second International Conference on Health and Design. Stockholm, Sweden: Svensk Byggtjanst, pp. 49–59.

Vidler, A. (1990) *Claude-Nicolas Ledoux*. Cambridge, MA: The IT Press.

Williams, C. (2009) Lawyers Scared of Computers [Homepage of The Register], 23 December, [online]. www.theregister.co.uk/2009/12/23/cps_paper/ [accessed 24 December 2009].

Williams, J. (2012) *Zero Carbon Homes: A Road Map*. Oxon, UK: Earthscan from Routledge.

Worthington, J. (2005) *A Future for Architectural Education in Ireland*. London: DEGW.

Young, N.W.J., Jones, S.A. and Bernstein, H.M. (2008) *SmartMarket Report on Building Information Modelling – Transforming Design and Construction to Achieve Greater Industry Production*, USA: McGraw Hill Construction.

7 BIM for all or some
Issues facing educators and practice

7.1 Separating discipline-specific from cross-disciplinary activities

The processes utilised during the design and construction of a building follow a cycle largely dictated by the contributions of the professionals involved, be they architects, surveyors or contractors. The contributions sequence is in turn determined by the type of procurement route chosen and associated contract, although some fundamentals will always remain unchanged, from design to procurement and from construction to use. Each contribution is completed by a specialist discipline or profession, each of which has its own specific processes and activities. For instance, an architect will go through a series of design stages, including consultations with a client, before a brief and a design reach a level where they can be shared with and contributed to by other disciplines. These specific activities are a reflection of an education system in the built environment sector which has historically focused only on the discipline (both in knowledge and skills) and, until recently, cross-disciplinary working has largely been left to be acquired in the workplace. A built environment graduate, a quantity surveyor for example, will have very little knowledge or experience of design activities and processes, which could make cross-disciplinary collaborative working a difficult task, at least to start with.

A cornerstone of a successful implementation of BIM is interdisciplinary collaborative teamworking from an early stage of a building life cycle, and all the benefits this collaboration is expected to add to both product and process, as will be fully discussed in Section 9 (BIM and teams). Key to the success of this new working environment would inevitably be to have the ability to separate discipline-specific from cross-disciplinary activities and processes. In a dynamic working environment in which the exchange of information is continuous and associated activities often overlap, being able to clearly distinguish between these two types of activities is critical to both the integrity of each professional discipline as well as to the running of cross-disciplinary processes. The direct impact on each discipline would be to identify and develop BIM-compatible workflows, templates and protocols that will govern the specific and shared activities as well as remove any ambiguity, be it professional, legal or otherwise. For instance, an architect ought to have a workflow that determines the specific activities involved in the overall design process, such as type, level and timing of information to be shared with other disciplines, protocols of information exchange including any data safeguards. Being clear and specific about the nature of activities involved will also help identify the resources and level of skills needed. This could prove to be a useful exercise for small SMEs whose contributions may be very limited, such as a small plumbing subcontractor whose activities may be restricted to accessing operational information and possibly providing feedback, as compared to an MEP (mechanical, electrical, plumbing) engineer, whose input includes design information, plant specification data, performance data, etc. Such clarity would also safeguard the integrity and interests of the disciplines and encourage true dynamic collaboration through which all contributions are well defined and understood; and, most importantly, lead to efficient delivery processes.

If the premise of being able to separate the discipline-specific from the cross-disciplinary activities is essential to a successful BIM collaborative working environment, which we believe it is, then a fundamental shift in how the disciplines are taught becomes a prerequisite. A shift that fosters a deep understanding of the discipline-specific activities and processes coupled with a practical working knowledge of the cross-disciplinary. This echoes Richard Saxon 2013, UK Government BIM Ambassador for Growth, who argued that BIM can't be taught with no read across disciplines and suggested, for example, 'crash weekends' where students from different disciplines work together solidly on the same multidisciplinary project. He also argued that reorganisations in higher education institutions (HEIs) could position construction and built environment undergraduates away from each other with unrelated disciplines, and could 'deepen' silos, which would not be helpful to learning what different professions do and how to work across such boundaries, deemed fundamental to developing the kind of professional necessary to make BIM a success.

7.2 Building strategies for collaborative use of ICTs

A latent and lingering question central to the BIM debate is, quite simply: can the industry cope, at least in the medium term? In migrating from 2D CAD to BIM, it seems reasonable to suggest that UK construction will travel through a transition phase in relation to current work practices and may challenge the capacity of education and training to support change in the workplace. There are signs that this transition phase will lead to two distinct groupings: a minority of BIM early adopters, predominantly made up of large design/construction organisations, will take the lead, followed by a majority of predominantly SMEs and micros with a widely variable Level 2 BIM implementation. It could be argued that this divide already exists. A number of key issues need to be seriously considered, particularly for the latter grouping. First, the timescale: the 2016 Government-mandated deadline is very short given the practical and cultural issues involved. Second, the fast-moving pace of ICTs, in particular the constantly evolving BIM systems software and the inevitability of change, will impact on any organisation trying to introduce and implement a new business model. And third, the human, financial and expertise resources needed to effect change may put a considerable strain on the industry, in particular on micro-SMEs. In addition, the ongoing economic recession, which is predicted to continue beyond the 2016 UK Government BIM deadline, will continue to impact heavily on the capacity of the built environment professions to cope with significant and tangible change.

Mervyn Richards, in his book *Building Information Management: A Standard Framework and Guide to BS 1192* (Richards 2010), suggested that key to addressing these perceived challenges is the UK industry's will, impetus and agility to migrate from document-dominated to information-centric digital environments; the latter with the potential to harness the power of information technologies to facilitate greater efficiencies in the delivery of whole life data for buildings and infrastructure. But, as Bronwyn and Khan (2002) have argued, the contribution of technological innovation to economic growth can only be realised when and if new technologies are widely diffused and used; and in the case of the construction industry, that means among the 80 per cent of SMEs and micros. Also, as noted in Robinson's paper (2009), 'A Summary of Diffusion of Innovations' in *Enabling Change*, technological evolution predicated

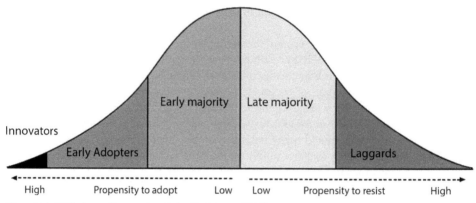

Figure 7.1 Diffusion of innovation (after Robinson 2009)

on change to work practices and cultures may diffuse at different rates among populations and peer groups, as illustrated in Figure 7.1.

Answers to the above questions lay in a gradual and incremental migration to BIM over a period of time through a combination of push and pull factors. Working with BIM early adopters either as partners or subcontractors, SMEs would gradually become familiar with BIM workflows and develop confidence and experience in their specialist areas of work. The latter combined with investment over time in BIM technology infrastructure and expertise, will lead to a full migration. However, the migration timescale will always remain dependent on each organisation's characteristics, such as size, scale and complexity of projects, clients' BIM demand (private or public), business aspirations, etc. Klaschka (2014), in *BIM in Small Practices,* recommends that an SME allows itself a period of one year to get to grips with BIM software and associated workflows and, most importantly, to reach a point where it can derive real business value from full implementation.

A successful migration of the construction industry to Level 2 BIM and beyond can only be sustained in the long term by a supply of skilled graduates. In recent years, many undergraduate courses have extended their programmes to include 3D modelling, environmental analysis and some aspects of BIM. The latter would naturally be expected to feature more prominently in the curriculum throughout built environment courses in anticipation of the 2016 Government deadline. On that point, Rosenbloom holds the view that developing projects using data-rich architectural software packages is not in itself BIM. Being able to demonstrate evidence of collaboration across disciplines in developing workflows is an essential ingredient of the mix.

Currently, it could be argued that there is a lack of clear educational standards for BIM despite a number of recent initiatives, such as the UK CIC BIM protocol, targeted at built environment professionals and their clients. The BIM Academic Forum (BAF, a grouping of a number of academics from UK universities), in conjunction with the Higher Education Academy (HEA), produced a document outlining proposed level learning outcomes for BIM education and training. In addition, a number of master's courses dealing directly or indirectly with BIM are being offered together with a plethora of short courses and CPD events provided by software houses, professional bodies and other software agencies. In this context, the Australian experience, as summarised in Rooney's *BIM Education – Global Summary Report – 2013,* provides an insight and highlights the key obstacles. A national BIM working group of industry and academic members concluded that introducing BIM into academia could be a difficult change process and highlighted a range of challenges. First are the difficulties of introducing a perceived new topic into an already crowded curriculum; second is the unfamiliarity of academics with the practicalities of BIM and workflows of related technologies. The third challenge is the resistance among some teachers to alter established teaching methods, and fourth is the inability to bridge the traditional educational silos of Architecture Engineering and Construction (AEC) and deliver collaborative undergraduate built environment programmes.

All the above initiatives seem to lack a solid point of reference in terms of clear benchmark standards or national guidelines on which academic and training programmes could be based. This lack of clarity has the potential to create confusion amongst practitioners and the industry as a whole in identifying the skills to meet the 2016 challenge and beyond. A way forward would be a partnership of education institutions including QAA, the construction industry and professional bodies. Such an alliance could provide clarity of curriculum, guarantee relevance to practice and ensure recognition of standards and qualifications. A case in point is the emerging role of the BIM manager.

Education for the Built Environment (E4BE) was created to fulfil the partnership role similar to the one described above as a tripartite and strategic grouping representing professional bodies, Higher Education (HE) and employer interests through CIC and Construction Industry Training Board (CITB). The alliance was tasked with acting on priorities impacting on the built environment in response to the UK Government's 2025 strategy. E4BE's engagement is intended to help shape capability and build capacity in the construction sector by enabling:

- professional institutions to extend their engagement with higher education in collaboration with other stakeholders, so that value is added to what higher education offers to industry;
- higher education providers to meet the needs of industry by working with other stakeholders in a more integrated way to satisfy higher education needs given constraints in the sector; and
- employers to address the need to skill new entrants (including employed, mature entrants and job changers, as well as new young entrants), and reskill/upskill/'recalibrate' mature entrants and the existing workforce.

To what extent E4BE's evolution and interventions may be influential in disseminating BIM practice across the broad spectrum of UK construction interests remains unclear at this stage. An effective engagement with built environment professional bodies seems to be missing so far. Without this influential link to professional courses (through their accreditation muscle) and education institutions, any outcomes would be limited in embedding BIM ethos and the practical skills that would be desperately sought in the next few years.

The collaborative use of ICT, by SMEs and micros in particular, requires the acknowledgement that a number of fundamental issues need to be overcome first as part of any transition strategy leading to Level 2 BIM. Collaborative working via a central BIM system can be an alien means of communication to more traditional, less efficient practices used to working with traditional workflows.

Adopting collaboration tools needs buy-in from all involved. BS1192 requires participants to agree a method of recording each issue and receipt of digital data, and also agree what constitutes an acceptable transfer. This active engagement may become a burden on small organisations to start with, and early recognition and planning would ease the transition.

The collaborative approach to design and construction with newer contract types has become much more important. Under recent contract types, such as design and build, the consultants actually undertake much less of the base design than they would have done in the days of 'traditional' contracting. Increasingly, it is much more likely that the supplier or specialist subcontractor will undertake the detail design with only overall co-ordination required by the consultants. This change, an opportunity for SMEs, requires much more efficient information exchange to allow speedy and efficient design to take place without people working with outdated information. Recognising these major industry changes and putting in place mechanisms in the transition strategy could help SMEs capitalise on new opportunities and adapt to new working methods.

Collaboration systems themselves are not foolproof and the use of these systems will require diligent uploading, downloading and document checking. Only this will ensure that there are no issues related to the use of outdated or incorrect information and this requires the users to have an efficient document management and control procedure.

7.3 Applying need-to-know principles for thinking and doing to pedagogy and practice

The changes that have occurred in the construction industry over the last decade or so, and that are still ongoing, in relation to the overall digitisation of the industry – including the major shifts in asset data management and exchange, innovative working processes and BIM, amongst others – lead the observer to believe that these paradigm shifts have the hallmarks of a 'revolutionary' change. These major changes have affected all aspects of the industry in a relatively short space of time. Assimilating all these changes, making a working sense of them and applying them to their business environment could look, legitimately perhaps, a daunting task for most built environment organisations, and particularly SMEs and micros.

Making sense of this major industry shift and, most importantly, extracting maximum advantage may lie in an evolutionary approach rather than the perceived revolutionary one. A step-by-step approach based on three fundamental criteria could help in this direction. First, is having a detailed understanding of the organisation's business structure, working methods, resources and overall capabilities, which would enable strategic judgement at all levels to be made. Second, is having a grasp of the changes that will affect all aspects of the business and their real impact in the short, medium and long term in relation to the above industry shifts. And third, is making sound priorities to implement change by selecting areas of the business on a need-to-know basis.

The need-to-know principle is a pertinent one for both pedagogy and practice within the context of the evolutionary approach to BIM implementation. An SME will easily become overwhelmed should it attempt to introduce wholesale changes to all areas of the business simultaneously. However, a careful analysis of the workflows and tasks, followed by an identification of the precise knowledge and practices needed for those tasks, could be manageable and any associated risks contained, with little or no impact on the day-to-day running of the practice. A small design practice, for instance, wishing to transfer from 2D drafting to a 3D data-rich model (pre-Level 2 BIM), may elect a small project, say a house or an extension, and allocate a limited resource, such as a new graduate familiar with 3D modelling, and use the experience as a test bed for a managed full transition. Such an experiment would help build confidence and capability within the practice, identify gaps to be filled, and may also be used as training for other members of the practice unfamiliar with 3D modelling and BIM processes. There is growing evidence that similar strategies are being developed and implemented by micro-SMEs, indicating self-awareness and initiative.

In a recently documented example, a micro-SME design practice devised a self-help strategy, initially involving one afternoon per week for two technicians to familiarise themselves with the Revit BIM authoring package and develop skills in 3D project information production. That exercise ran in parallel with live projects, which continued to be developed using 2D CAD. The goal was twofold: to gauge the feasibility of introducing 3D workflows to the practice and to run an early comparative analysis exercise between 2D and 3D working in relation to some key performance indicators, namely potential practice capabilities and identify efficiencies that may be pursued further. In this particular example, the information database potential of the BIM authoring package was not tapped significantly by the practice but, for example, two previously discrete workflows were harmonised by embodying predictive energy modelling and analysis input/output into the BIM model using Autodesk Green Building Studio and the ubiquitous gbXML file protocol. Plus of course, in outlining the formative stages of migration towards Level 2 BIM, the practice has not yet tackled the key process protocols, in particular PAS 1192-2 and the underpinning BS1192:2007. But it is a start, a short but useful record in the raw of a micro organisation's will to embody change and embark on a BIM journey, and a precursor to a more fundamental review into how data is generated and managed in a holistic way within the organisation.

It is interesting to compare this self-help approach with the idea of 'BIM training' being offered by commercial organisations, which might represent a significant on-cost for a small organisation. It is also interesting, in noting that BIM4SMEs online training packages have received limited take-up by the industry, to pose the hypothetical question: should migration to Level 2 BIM be within the capabilities of even the smallest organisation?

Similar to practice, the need-to-know principle could also be applied to the pedagogy of interdisciplinary collaborative learning. By extending the concept of separating discipline-specific from cross-disciplinary activities (see Section 7.1 above), each discipline ought to be able to identify its specific collaborative knowledge and skills and embed them in its education programmes at the appropriate level. A parallel exercise should also be applied to those cross-disciplinary needs for both simulated and applied collaboration, which can only be achieved effectively through a partnership between educational institutions and the industry. In this fashion, the need-to-know principle can assist in finding a clear and focused pedagogy of learning away from a learn-all culture engendered by competing interests, some legitimate and others less so. This is illustrated in the plethora of software tools constantly flooding the market, all claiming collaborative solutions; this, coupled with the lack of established standards, makes the pedagogical choices difficult. Within this context, making those choices on a need-to-know basis must be closely linked to the discipline-specific and cross-disciplinary appropriate requirements.

7.4 New knowledge through processes of enquiry and experimentation: the role of professional bodies and industry umbrella organisations

Undoubtedly, the transition to Level 2 BIM and beyond will vary from one discipline to another and also between organisations depending on a host of reasons and rationale; some may reside in company capabilities, size, ambition or simply market demand (see Section 7.2 above). Market forces will certainly play a role in how and when BIM is adopted. However, at the core of this transition is a process of acquiring new knowledge, making good sense of that knowledge then adapting and implementing it to suit the various business processes and workflows of the organisation regardless of any other consideration. Drawing parallels with the introduction of CAD in the 1970s, BIM knowledge is dissimilar to that of CAD in that the transition from drawing board to the screen involved a change of medium but the rules governing buildings' graphical representation remained largely unchanged. Once the software commands and functionality are learned, the rest is business as usual. Learning to work with BIM goes beyond graphical representation, digitally modelling a real building rather than drawing it using established norms of graphical representation, and involves first an understanding of how buildings work, perform and are built, and second, having a grasp of the basic processes, at least, of how the design and construction of a building is contributed to by other team members including the various interactions involved.

This paradigm shift, as often referred to in the BIM jargon, continues to evolve in terms of knowledge and application and the construction industry still has a fair distance to travel before working methods are tested and become established. This evolving knowledge could, we would argue, come about through structured processes of enquiry and experimentation facilitated by professional bodies and industry umbrella organisations. A glance at the specialist press and the weekly event schedules quickly reveals the plethora of BIM events taking place at local through to international level in the form of conferences, CPD sessions, seminars, etc. Despite the wealth of debate, one questions the existence of a co-ordinated approach based on common goals and clear industry guidelines. Arguably, here lies a major opportunity for professional bodies and industry agencies, together with HEIs, to map a collaborative common framework linking in an effective manner education and training to professional practice through maintaining connections between undergraduate education, research and professional practice. Industry studies have consistently demonstrated a dearth of R&D in construction. One key facet of an evolutionary paradigm for BIM uptake is that it should follow a reasoned, evidence-based and consensual pathway towards implementation. Collaboration across disciplines and the involvement of clients are thought to be key to achieving that objective. At the 2013 E4BE BIM workshop in November 2013, CIC Chief Executive Graham Watts emphasised how fundamental BIM is to the UK construction industry's future, and how critical higher education is in ensuring BIM is developed and delivered to assist in hitting strategic industry targets. The impact of experimentation and applied research in BIM implementation can be most effective when carried out within a professional environment, with a well-structured and guided approach. A number of key action points may be considered to assist acquiring knowledge through the processes of enquiry and experimentation:

- professional bodies' effective engagement through professional courses by, for example, including BIM learning outcomes in their professional courses mapping;
- an effective co-ordination of practical training initiatives at different levels between industry umbrella organisations, HEIs and professional bodies;
- the encouragement of enquiry and experimentation through industry and education linkages similar to recent initiatives such as BIM academies;
- sharing of practical information through real case studies to support SMEs and micros who may lack capacity in this area; and
- generally more investment by the industry in research and development.

References

Bronwyn, H.H. and Kahn, B. (2002) 'Adoption of New Technology', in: D.C. Jones, ed., *New Economy Handbook*, Cambridge: Academic Press, [online]. http://eml.berkeley.edu/~bhhall/papers/HallKhan03%20diffusion.pdf [accessed 19 January 2015].

Klaschka, R. (2014) *BIM in Small Practices: Illustrated Case Studies*. UK: NBS Publications.

Richards, M. (2010) *Building Information Management: A Standard Framework and Guide to BS1192*. London: BSI Group.

Robinson, L. (2009) 'A Summary of Diffusion of Innovations', in *Enabling Change*, [online]. www.enablingchange.com.au/Summary_Diffusion_Theory.pdf [accessed 19 October 2014].

Rooney, K. (2014) *BIM Education – Global Summary Report – 2013*, Construction Information Systems Ltd, Sydney, [online]. http://bim.natspec.org/ [accessed 19 October 2014].

Saxon, G. R. (2013) Growth Through BIM Report, [online]. http://cic.org.uk/publications/ [accessed 2 October 2015].

8 Challenges, risks and benefits for SMEs

8.1 Same old, same old: the need to modify work practices and cultures

The transition to a BIM collaborative working environment poses a number of challenges and risks, as well as benefits, for small and medium-sized enterprises in the construction sector as a whole. The technology gap within the construction industry continues to widen between large companies and SMEs/micro-SMEs. Amongst the former, ICT is pervasive and has become a key infrastructure covering all aspects of business, including design, construction, asset management, life cycle, marketing, cost management, etc. Amongst the latter, however, ICT is often limited to traditional 2D drawings, perhaps some static 3D visuals, email and possibly a symbolic internet presence. The status quo creates a divide in the industry and potentially may lead to a macro versus micro effect that will certainly hamper any effort for collaboration within the industry. The real challenge here, one would argue, is for those large organisations with experience and expertise in BIM to match their legitimate marketing discourse with a similar level of actions by providing reliable and credible case studies and even collaborate with SMEs to raise industry level and help to achieve a faster and smoother transition to BIM.

Access to BIM by SMEs and in particular micro-SMEs tends to be hampered by a lack of ICT infrastructure as an enabler to a quick deployment of BIM. Working in a dynamic workflow is a major shift in working practices and extends beyond acquiring expensive hardware and software. A radical change of attitude and work practices is required to work in this dynamic and collaborative environment. Furthermore, the learning curve is lengthy and demands sustained training and upskilling of personnel, as well as changes to the organisation's business model. A recent study by the authors examining BIM as a collaborative tool concluded that, contrary to literature claims, a case study has shown that the present investment, in terms of time, cost and effort required to implement the technology means that BIM is unlikely to be adopted on small, simple projects or by micro-SMEs where conventional CAD is adequate.

The burden of the additional expenditure is not insignificant in a highly competitive and difficult economic climate in which small businesses are often struggling to stay afloat. The reported cost of installing a full BIM station varies from £10k to £1k at the lower end and from £30k to £70k at the leading edge end. And in the absence of a clear and coherent national/professional strategy on BIM on the one hand and the lack of clarity on the level and timescale of return on investment on the other, it is even more difficult to make substantial financial commitments, as highlighted by BIM4SME:

Moreover, as the group are SMEs themselves the equation of cost is foremost in our minds, but perhaps the more important question is 'what do I get for my money, what are the benefits and how long before my investment is paid off'.

(BIM4SME 2015)

Embarking on a BIM implementation process is not only a daunting task but carries considerable risks from a business viewpoint. SMEs do not possess the financial resilience of larger organisations to be able to absorb potential losses or dedicate significant resources, be they financial, human or time. Without very careful planning, getting it wrong could mean business ruin. It is these risks and challenges that call for careful planning and gradual implementation of collaborative BIM together with strategic and technical support for SMEs.

Despite the above risks and challenges, we strongly believe that a managed transition to Level 2 BIM and beyond offers sound benefits for SMEs from both a pragmatic business viewpoint and, most importantly, their long-term readiness and resilience for a rapidly changing construction industry. Work efficiency benefits are well documented in the literature and include access to up-to-date information, error reduction/elimination, ability to visualise complex information, avoiding duplication of work and rework, etc. Other more strategic business benefits may include access to bigger markets, partnerships with larger companies, expansion to emerging niche markets, and the ability to compete using the same tools and skills as larger organisations.

Level 2 on the BIM Maturity Index (Figure 8.1) assumes that everybody ought to be operating at the very top end by 2016. What is the level of progress towards this goal? The reality is a mere guess, particularly amongst the majority of small companies across all built environment professions. It remains difficult to build a full picture of the level of true understanding

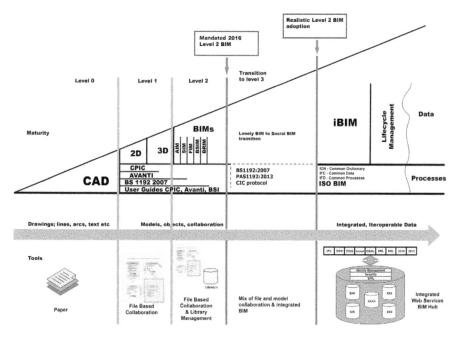

Figure 8.1 Modified BIM Maturity Index: from mandated to realistic BIM adoption through a transition period (after Bew and Richards 2013)

and implementation due to the lack of reliable data and limited take-up of BIM surveys. Recent surveys tend to be unrepresentative due to the small number of respondents. For instance, the National Building Specification's (NBS 2012) survey received 1,500 respondents in contrast to 30,000 RIBA members. Despite the low intake in these surveys, the results do not provide an encouraging reading; the National Federation of Builders' (NFB 2012) survey provides an insight into the state of readiness of construction SMEs in relation to Level 2 BIM uptake. Of all contractors surveyed, only 30 per cent work with electronic 3D drawings, 54 per cent of which are large contractors and only 25 per cent are SME contractors.

Some might also argue that a good proportion of the above claims are file sharing through a management system, often set up by the main contractor or the local authority and used by the various professionals, or simply through a simple dropbox. Only a number (but increasingly so) of large design and construction organisations run their own complete data management systems. The type of data shared tends to be mostly 2D drawing or PDF files, spreadsheets and Word files. One may even argue that most of these repositories are there to comply with BS1192:2007 and at best provide a level of passive data exchange, which may assist co-ordination.

Achieving Level 2 BIM by the Government 2016 deadline may not be within the reach of all, and real change requires a modification of work practices and cultures. The following modifications are essential for any small business aspiring to enter the digital age and harvest some of its rewards:

- It is a recognised fact that in the new digital age there is a generational skills gap within the construction industry. First, a new, younger generation of graduate professionals, well-trained and versed in ICT tools and working methods, has been entering the construction industry professions next to a well-established older generation, whose knowledge of these technologies is limited to say the least. And often they do not share the same attitudes as far as the potential of these technologies. Second, the managerial power within the industry predominantly resides with the older generation. The latter is often sceptical and reluctant to adopt unfamiliar technologies and working practices in which the risk is perceived to be high. A resolution of this generational gap is critical through a combination of faith in the technologically skilled younger generation combined with an element of business risk; and the recognition that the status quo is not an option in the medium to long term.
- There seems to be a resistance amongst SMEs and micro-SMEs in particular to look beyond tools required to do the job. This sense of immediacy, often born out of work environment necessity, needs to give way to an attitude of looking beyond the immediate through continuous planning and investment, and as the best guarantor for long-term business success.
- Embracing BIM technologies within the context of digital Britain does not involve a series of technical fixes, but demands first a belief in this radical change and second the consistent embedding of these new technologies in the business processes and workflows of the whole organisation.
- Developing a business model and outlook that recognise the inevitability of change and continuous learning both at the corporate and individual level is essential. Once a full commitment is made, it could be argued that real transitional change to BIM for an SME may be implemented more efficiently and in a shorter timescale. Smaller staff numbers, versatility of skills and relatively smaller number of workflows, compared to larger organisations, could render the change process more manageable.

•   The inherent agility of SMEs to adapt to changing business conditions could be deployed to modify work practices and shorten the transition period to Level 2 BIM and beyond.

8.2 Keeping up with changing technologies, opportunities and risks

It is generally acknowledged that in the digital age there is a generational skills gap within the construction industry. First, a new, young generation of graduate professionals, well-trained and versed in ICT tools and working methods, has been entering the construction industry professions next to a well-established older generation whose knowledge of these technologies may be limited. Also for micro-SMEs in particular, there may be a resistance to looking beyond tools required to do the job, particularly in a difficult economic climate where business survival becomes the overriding priority. First small, medium-sized and large organisations may not share the same attitudes and values regarding the potential of these technologies. Second, the managerial power within the industry predominantly resides with the older generation, who are often sceptical of, and reluctant to adopt unfamiliar technologies and working practices, which are perceived to embody high risks. Clearly, the industry is going through a major transition and the BIM Task Group's push-pull strategy may help to firm up minds, shorten the transition period and open the doors wider for the aspirations of younger professionals. Opportunities as well as risks certainly exist, but the strategies that are put in place and the actions that are taken will be the secret for success or failure. One thing is certain: a gradual implementation of these new technologies can meet the transformative and evolutionary effect only if it is matched with a strategic top-down business vision.

Opportunities offered to SMEs and micros by BIM technologies have been the subject of a long debate and may range from work efficiencies to a full business transformation. The American experience, as advocated by the American Institute of Architects (AIA), highlighted the significant access to business opportunities by small and medium architecture practices in the USA due to their increased competitive BIM capabilities on larger projects. These capabilities may be focused on specialist expertise such as environmental modelling, integrated project management and advanced visualisation. Providing these additional services and others, traditionally seen as either costly or perceived as being of little value, not only improves customer service but also increases business opportunity. There is agreement within the industry that significant work efficiencies can be made to workflows, overall management of the business, project data management and communication, amongst others. BIM technology, if fully implemented, can bring a small business's work practices and skills to a level comparable to that of a larger organisation, hence facilitating business opportunities through, for example, subcontracting and direct competition for work packages or large projects via partnership. The transition to full implementation of BIM has already begun, creating new business opportunities as well as job roles which may be compared to those created by the introduction of CAD in the 1970s and 1980s. BIM manager, design manager and information or data manager roles are not uncommon in the industry. These roles require an expertise in the application of these new technologies and, most importantly, are oriented to a newly qualified young workforce. New niche businesses specialising in 3D modelling, data management, advanced building modelling, environmental modelling, etc. are an opportunity for new young business starters and an opportunity for expansion for SMEs.

Any opportunities created by the introduction of BIM technologies, however, are not free of risks, and smaller organisations are more exposed given their limited infrastructure and ability to sustain sudden change. One may argue that the biggest risk facing SMEs is no change at all and, to quote the American architect Thom Mayne during the AIA Building Information

Modelling Panel Discussion, 'If you want to survive, you're going to change; if you don't, you're going to perish. It's as simple as that' (Mayne 2005). It is, therefore, a question first of identifying the changes required; second, of setting up a change implementation plan; and third, of finding the change management mechanisms appropriate to the business.

Changes would principally include updating hardware/software, training staff, developing a BIM business ethos, updating or introducing BIM workflows, and tackling old attitudes and work practices. Any change plan for a small construction organisation ought to be evolutionary and follow clearly identified incremental steps, taking into consideration business decisions on key issues such as financial commitments, staff workload, time to embed changes for the long term and the ability to add value to the overall business. The appropriate change management mechanism for an SME may not be a standard 'off-the-shelf solution', but rather one that is based on the company's history, market, client base and, most importantly, staff resources. For instance, a change mechanism may be based on partnerships with other construction SMEs whereby shared experiences and expertise are of mutual benefit. Experimentations by micro-SMEs embarking on a BIM implementation journey are not uncommon, and limited evidence suggests that the will to do of organisations may be another key enabler, a view advocated by Mervyn Richards.

8.3 Tailoring BIM to suit business models: filtering out achievable BIM outcomes in relation to organisational size and resources

Within the UK construction sector, BIM literature (the evangelical model as opposed to the evolutionary one), with all its facets and mantras (collaboration, communication, project efficiency, carbon reduction, whole life asset management, etc.), has focused on implementation in large design and construction companies. These organisations operate at a much larger business scale than SMEs and micros, who represent the majority interests in the sector. For these large organisations, engaging with BIM may offer competitive advantages, which can be easily afforded, not only to maintain leadership in the market but also to harvest the business benefits BIM may bring to the table. In that context, where does the debate leave the SMEs in the sector? Where are the argument and counter argument necessary to feed informed decision-making, particularly for small companies? It seems that until recently, this 90 per cent majority stakeholder interest has been left on the margins of the debate. With the cut-off date for the UK Government's mandate imminent, will there be a gradual awakening, realisation and actions with respect to how BIM may impact on UK construction in the round (the evolutionary model)? From the subgroups set up to deliver on the Government's BIM agenda, BIM4SME has developed as a cross-disciplinary grouping of interests championing BIM and promoting, in particular, the interests of construction sector SMEs. Its primary and only focus is to support SMEs in their understanding and use of BIM, whether they are consultants, contractors, specialists, suppliers or manufacturers.

As BIM paradigms continue to emerge, develop and evolve across construction disciplines, the idea of BIM requiring new business models has become more established and is challenging traditional methods of delivering building projects. Typically, in a traditional model, the overall process consists of two interlocking subprocesses or activity nodes, design/construction activities (process node) and policy and codes (policy node). Technology is normally embedded within the activities of each node, with limited crossover. A typical example would be an architecture design office running specialist software on free-standing PCs or linked via an intranet and all communications with organisations within the policy node are done through traditional channels of communication, such as meetings, post and email.

Razvi, in an article entitled 'BIM and the Process Improvement Movement' published in the BIM Think Space blog (2008) argues that, in a BIM business model, a third technology node which interlocks with the other two has become critical to process development, as illustrated in Figure 8.2. In a BIM working environment, the interlocking area between the three nodes will proportionately increase with the maturity of the BIM system, providing more opportunities for collaboration between stakeholders.

The IT infrastructure and expertise required to support a BIM model are sufficiently complex that often they need to be run and managed by external specialist network and data management agencies. A number of these companies are already active in the market, including data repositories and providers of web and cloud-based construction collaboration technologies. Such a business model could be considered imperative for an organisation in the construction sector to capitalise on the benefits of implementing BIM (efficient workflow, collaborative working, building partnerships, good communication, etc.) and add value to their business.

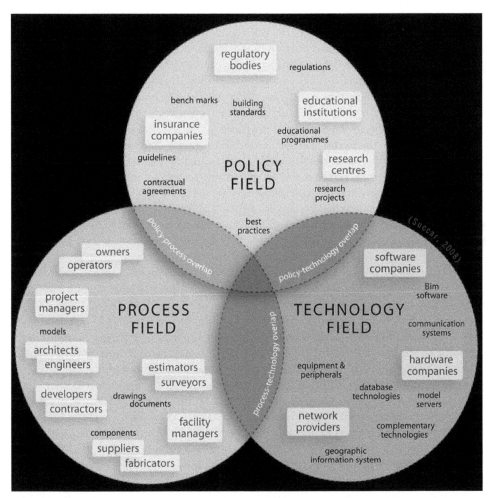

Figure 8.2 The three interlocking nodes of BIM: in a BIM working environment the common areas will be much bigger, providing increased collaboration opportunities between all stakeholders (after Razvi 2008)

SaaS provided by technology companies has passed the good idea stage and its uptake by construction industry organisations is on the increase. This method of accessing advanced ICTs may help to free design and construction organisations from a burden they neither have expertise in nor have as a core of their business. Additionally, the rapid change in IT software (release of new versions and issues of compatibility, amongst others) may make leasing BIM and data management software an attractive proposition that allows organisations to focus on areas of expertise and business growth.

8.4 Considering generic and bespoke aspects of BIM principles and practice

'A comprehensive and growing collection of BIM (Building Information Modelling) objects spanning all building fabric systems'. This is how the NBS describes its National BIM library as an example of generic BIM. Modelling buildings' performance, constructability, health and safety and so forth rely on populating the BIM model early in the design process to facilitate testing, simulations and early problem resolution. This is expected to happen before many consultants have worked the details of their design. Generic objects developed on standard, average components are often used early on in the design to be able to run these simulations and test performance. Increasingly, manufacturers are supplying the market with their specific component and material parametric objects ready to be used in a BIM model. The use of these objects is common practice in many software packages. A simple, typical example would be a wall in a Revit model for a standard family home, which would consist of three layers: internal, external and middle, for insulation and structure.

Three features are key to the usability of the NBS national database, namely integration, intuitiveness and quality. Access to BIM objects is facilitated by a direct link from the design tool using bespoke plug-ins for the major design software packages. Intuitiveness is embedded in intelligent search and filtering capabilities following the needs and established practices of specifiers and, most importantly, the quality of data as all objects, including contained data, need the highest level of accuracy and standards. The generic aspects of BIM, whether a component database or BIM model viewers, contribute to the implementation of BIM and can be easy access options to SMEs and micros embarking on a BIM journey.

With reference to Section 8.3 above, the number of technology companies already active or entering the market is increasing. These companies provide bespoke technological solutions ranging from complete BIM solutions to very specialist service applications. Clash detection and specialist project document management systems are typical bespoke solutions. For instance, BIMsync from BIM Technologies, in beta test at the time of writing, is an automated specification tool allowing for two-way syncing between specification and model production of a validated approved specification. It is an innovative application that is exploiting a developing niche BIM market, bridging the gap between a BIM model, specification and approved validation of specification. At the other end of the spectrum, Solibri is a bespoke model checker software with wide capabilities. It performs advanced clash detection, deficiency detection, BIM and accessibility compliance, model comparisons, etc. at the higher end of BIM workflow.

Navigating generic and bespoke aspects of BIM is an exercise every built environment organisation will have to go through as part of an analysis of their business model and the separation between discipline-specific and cross-disciplinary activities. The analysis exercise would help frame business and workflow priorities and identify aspects of BIM principles and tools relevant to the business model.

References

Bew, M. and Richards, M. (2013) BIM Maturity Index Diagram ©, available at 'BIM Working Group report', [online]. www.bimtaskgroup.org/wp-content/uploads/2012/03/BIS-BIM-strategy-Report. pdf [accessed 10 May 2015].

BIM4SME (2015) [online]. www.bim4sme.org/about/the-organisation/ [accessed 10 May 2015].

Mayne, T. (2005) The Building Information Modelling Panel Discussion, at the 2005 AIA National Convention, Las Vegas.

NBS (2012) National BIM Report 2012, National Building Specification, Newcastle. [online]. http://www.thenbs.com/pdfs/NBS-NationalBIMReport12.pdf [accessed 3 September 2013].

NFB (2012) BIM: ready or not?, National Federation of Builders, Crawley. [online]. http://www.builders.org.uk/resources/nfb/000/318/333/NFB_BIM_Survey_BIM-ready_or_not.pdf [accessed 3 September 2013].

Razvi, S. (2008) BIM and the Process Improvement Movement, Building a Case for a Combined BIM–CMMI Framework, [online]. http://changeagents.blogs.com/thinkspace/files/BIM_ThinkSpace_The_Process_Improvement_Movement_by_Sohail_Razvi.pdf [accessed 12 February 2015].

Solibri Model Checker (n.d.) [online]. www.solibri.com/ [accessed 12 February 2015].

Part III

Teamworking and information management

9 BIM and teams

9.1 The role of clients and their support organisations

Clients, nowadays, recognise the value of design integration and collaborative working and are increasingly insisting that their design-construct team have efficient collaboration amongst the different members. They also recognise that an efficient project collaboration tool will assist the development process and yield quality project information. A 3D model, compared to 2D drawings, helps clients to better visualise and understand design options at an early stage of the design process and hence increase the chances of more accurate feedback and quick decision-making. The use of non-parametric 3D models and animations has arguably become a near common practice, particularly at the early design stages; and evidence from practice seems to indicate that clients value this method of interface. However, there seems to be a mismatch between clients welcoming, even demanding, design and construction information in 3D format and not appreciating the value of embedded information (data) in those same drawings. Moving towards a digital working environment, this mismatch gap needs to be bridged to allow for clients to actively engage, recognise the real value of the project embedded data and be prepared to pay for it.

The role of clients was clearly articulated in the second strand of the Government Task Group push–pull strategy, which focused on the client 'pull' element and how the client should be very specific and consistent about what they specify when commissioning an asset. This includes the need to be specific on a set of information (data) to be provided by the supply chain to the client at specific times through the delivery and operational life of the asset. This would rely on the careful definition of what data deliverables would be needed and when, linking in the standards and specification processes. The expectation is that this data delivery would have the dual benefit of ensuring that, first, complete information sets are delivered on time, enabling commercial checks and handover information delivery, and second, consistent digital handover information is delivered, enabling access to the design, costs, carbon and performance of the asset.

The above puts clients and their supporting organisations at the heart of the design and construction process by, on the one hand, enabling them to request higher standards of product delivery and associated information, and, on the other hand, expecting better collaborative engagement in the process. These expectations demand a level of client maturity, supported by the consistent supply of information and timely decision-making that may often be available to clients of large assets, be they government agencies or corporate organisations.

9.2 Setting up and managing teams

Krygiel and Nies, in their book *Green BIM: Successful Sustainable Design with Building Information Modelling,* succinctly summarised the historical evolution of design and construction responsibility as follows:

> During the time of the Renaissance, architects like Brunelleschi were afforded the opportunity to operate as a master-builder. This paradigm allowed one person to hold complete knowledge of both the design and construction methodologies implemented in the building process, partially due to the relative simplicity of the building system at the time. Integration, so to speak, was automatic.
>
> (Krygiel and Nies 2008: 55)

The role of the master-builder split, after the Renaissance, into architect/designer and contractor roles; one became responsible for the design of the building and the other for its construction. At this point, co-ordination and team collaboration became necessary to achieve the expected building outcomes. Ever since, technologies have continued to evolve, and the design and construction of building has become increasingly complex. As a consequence, building systems have needed greater specialisations, which has led to the growth of numerous built environment professions (architects/designers, engineers, surveyors, project managers …). This has led to the compartmentalisation and fragmentation of the construction industry which in turn has led to a silo mentality.

Recent developments in the construction industry – since Latham's *Constructing the Team* report (1994) and the more recent drive by the British Government's Construction Task Group – have generated an unprecedented momentum that is changing practices and attitudes. Key to those changes is how design and construction teams work together using advanced information technology tools to manage the ever-increasing project information in an efficient and collaborative manner.

The complexity of construction projects and related processes requires the input of a design and construction multidisciplinary team to deliver project outcomes. The Royal Institute of British Architects, (RIBA) Plan of Work recognised the importance of collaborative teamworking in design and construction since the early 1960s. It recommends the setting up of a design team early in the design process (Stages 0 and 1). However, members of a design team represent different disciplines that are often fragmented and at times adversarial, as indicated in the 1994 Latham report. With the different disciplines come different ways of working, thinking and talking about design. The different disciplines also become involved at different stages of the design process. Despite the above, the design-construct team is expected to deal with the multiple layers involved in today's complex design layout (structure, project data, cooling, ventilation, heating, security, plumbing, lighting, etc.). Figure 9.1 summarises these complexities, each of which requires expertise to deal with a host of issues ranging from design, planning legislation, and structural and environmental standards, etc. The number of layers, the various processes involved and the complex interactions between them all suggest an integrated and collaborative teamworking approach to design and construction, just as they suggest the need for robust tools to manage these complexities.

The historical silo mentality in the built environment professions is well documented, and breaking professional barriers has been advocated by the industry itself and its critics for many years. Multidisciplinary team collaborative working is not new in the construction industry, but it can be argued that its extent has until recently been limited to the minimum required by a

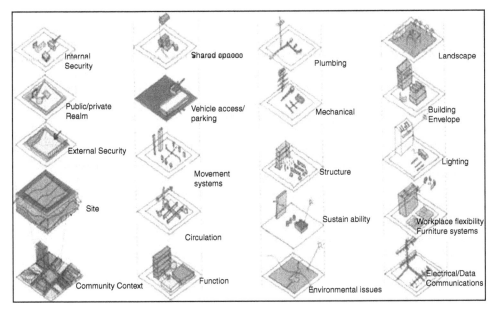

Figure 9.1 Layers of design (after Krygiel and Nies 2008)

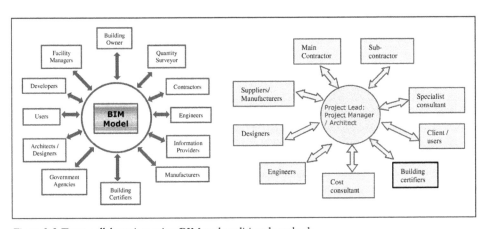

Figure 9.2 Team collaboration using BIM and traditional methods

project, often to resolve problems. The introduction and increasing implementation of BIM technology in the industry in recent years has significantly redefined design and construction team interactions and refocused the collaborative processes involved. Central to this redefinition has been a shift of focus from team members' roles and the contributions they make to the product itself, i.e. the building throughout its whole life cycle. This has not fundamentally altered the nature of team members' input, but it has created a collaborative working environment that fosters early involvement, transparent information exchange and reliable data process management and control as illustrated in Figure 9.2. As a result, the logistics of involving all stakeholders is made simpler and more easily manageable through a BIM workflow and therefore each member, be it a micro-SME, user group or individual, can actively participate in the collaborative process.

Exploiting the collaborative capabilities of BIM constitutes a cornerstone of the Government Construction Task Group push–pull strategy, which revolves around two key criteria: whole life cost (including savings in capital cost), and carbon performance (together with related processes). Within the 'push' element, BIM is seen as an enabler to timely and accurate information delivery, as well as to effective and transparent decision-making. This strategy renders the collaborative use of BIM by multidisciplinary teams, as opposed to BIM technical aspects, essential to a built environment practice or company of any size within the overall supply chain. At the same time, the IT tools, the technical aspects supporting any BIM system, are essential to the processes of design and construction team collaboration and management.

The collaborative approach to design and construction with newer contract types has become much more important. Under recent contract types, such as design and build, the consultants actually undertake much less of the base design than they would have done in the days of 'traditional' contracting. Traditionally the architect was the 'project administrator' and undertook all the architectural design with other consultants being responsible for structural, civil and services design, etc. Increasingly, it is much more likely that the supplier or specialist subcontractor will undertake the detail design, with only overall co-ordination required by the consultants. This change has required much more efficient information exchange, to allow speedy and efficient design and construction to take place without people working with out-of-date information. It was inevitable, therefore, that the leading clients, designers and contractors would welcome the implementation of collaboration software on their projects.

Numerous tools and systems are available, with web-based platforms being the most common. BIM systems linked through a network, intranet or extranet have emerged as robust tools with very high design and construction integration as well as collaborative working potential. Autodesk Revit and ArchiCAD are examples of a number of BIM authoring packages used in practice that are capable of providing a platform for collaborative working, data sharing and exchange, as well as open communication between team members and other stakeholders. However, collaboration utilising a BIM system needs buy-in from all team members, as if any party either chooses not to embrace the concept or does not use the tools effectively, then the system will probably fail to deliver to its potential. To more traditional and less technologically adept practitioners, BIM can be an alien means of collaboration and communication. It is critical to remember, however, that these collaboration systems are not foolproof, so their use will require diligence and timeliness in updating, downloading and uploading information. Only this diligence will ensure that there are no issues related to the use of outdated information and this requires each team member to adhere to an agreed efficient document control procedure.

Despite the evolving design-construct teamworking processes and the new collaborative tools being deployed to meet BIM Level 2 goals and beyond, a fundamental question still remains to be fully answered, namely, who will lead and, most critically, pay for the overheads related to the set-up and management of teams and associated systems? The anecdotal consensus in industry points to the client, being the main beneficiary of the final outcome both in terms of the facility being delivered efficiently and inheriting the model for life-cycle use. In practice, however, the picture is not at all clear; large design and construction organisations tend to absorb the costs for a combination of reasons that include, amongst others, market leadership, internal process efficiencies and self-promotion. The situation amongst SMEs and particularly micros remains ambiguous with isolated initiatives. An early consensus on design-construct team leadership, together with associated remuneration, would help expedite BIM implementation as well as harvest its benefits.

9.3 Addressing a range of paradigms: from lonely BIM to collaborative working

It would not be far-fetched to assume that the overwhelming majority of built environment organisations are currently using a combination of IT tools in their day-to-day technical and administrative business processes. The use of these tools can vary significantly, ranging from basic 2D CAD drawings to advanced 3D and BIM models. Although the construction industry is still heavily reliant on the use of 2D drawings, both paper and digital, in information production and operations stages in particular, the use of 3D drawings and animations has become common practice as an effective medium for communicating design intentions either to secure commissions or facilitate decision-making. Furthermore, the increasing use of parametric software platforms capable of producing data-rich design and construction information (increasingly known as BIM authoring software platforms), together with the rapid development in ICTs, such as the Internet, has brought about new working opportunities and changes to working practices unimaginable two decades ago. Implementing BIM working practices is still evolving as part of an inevitable transition period, which has given rise to a spectrum of paradigms ranging from lonely BIM to collaborative BIM.

Lonely BIM, also referred to as 'small BIM', is a term used to describe the practices of an organisation, project team or the whole market where the generated BIM models are not exchanged between project team members. Also, organisations at early stages of BIM implementation or who only generate mono-discipline models are considered to be practicing lonely BIM, a position between Level 1 and 2 on the BIM Maturity Index (see Section 8.1). Practicing this level of BIM can bring internal efficiency gains, build capability and develop BIM processes, a respectable position for an SME or micro organisation aspiring to the next stage, Level 2. The biggest prize for a lonely BIM practice, one would argue, is twofold: making the major paradigm shift from conventional building representation, using 2D drawing devoid of any information, to creating virtual models of a building, not only embedding valuable information related to performance, specifications, etc; and establishing BIM workflows and processes that help bring efficiencies and prepare for the next level of BIM.

Collaborative BIM, also known as 'social' or 'big BIM', is a term used to describe the practices of an organisation, project team or the whole market where multidisciplinary models/ BIM models are generated or are collaboratively exchanged between project participants. Collaborative BIM may also refer to elevated BIM maturity within a market. That is the mature Level 2 on the BIM maturity index mandated by the British Government by the 2016 deadline. This level of collaborative BIM is, at present, the privilege of a small minority of large organisations and remains aspirational for the majority of SMEs and micros in the UK construction industry, as argued elsewhere in this book. The benefits of collaborative BIM are well established in the literature, but the greatest value to a built environment practice is to endeavour to exploit the full potential of collaborative BIM beyond the statutory or contractual requirements of a project. Building partnerships and linked networks with suppliers, manufacturers, specialist consultants, etc. to foster long-term collaboration are typical examples.

9.4 Use of CDEs to manage information flow

To analogically paraphrase a politician's famous statement on the critical importance of the economy in politics, in Building Information Modelling the secret is in the 'I' (for Information)! There is general consensus that whatever acronym is used to describe BIM, with all its associated processes and workflows, information constitutes the most valuable asset for managing all activities related to a building life cycle. Building project data begins to be accumulated as early as the inception stage, then increases in complexity as the design and construction stages progress, to reach its fullest stage at handover stage, beyond which it is usually used for facilities management purposes. In a traditional workflow, the custodians of project data are the various members of the design-construct team, who generate data relevant to their contribution in the first place. Information generated from that data flows between team members in a linear fashion following project stages and only as required by established contractual arrangements and associated procedures. It is not uncommon that the information flow is not timely or complete, which tends to generate the notorious RFIs and cause delays, which in turn may impact on cost. Arguably, this reflects the silo mentality and fragmentation prevalent in the construction industry until very recently.

Common Data Environments (CDEs) are partly a remedy to the issues associated with the traditional information flow, and partly add efficiency to the management of those flows. A form of CDE (3D Environment) was pioneered in the early 2000s by BAA in the design and construction of Heathrow Terminal 5, which helped the subsequent development of these environments and related standards. A CDE is often defined as the single source of information, for a project, utilised to collect, manage and disseminate documentation, the graphical model and non-graphical data for the entire project team. This includes all project information whether created in a BIM environment or in a conventional data format. Creating this single source of information facilitates collaboration between project team members and helps avoid duplication and mistakes.

Uptake of these essentially information management systems intensified as a response to British Standard BS1192:2007, 'Collaborative Production of Architectural, Engineering and Construction Information', as illustrated in Figure 9.3. The standard established the methodology for managing the production, distribution and quality of construction information, including that generated by CAD systems, using a disciplined process for collaboration and a specified naming policy. This was a significant step forward in providing a unified framework for construction information management flow between all parties involved in a project. As a significant step forward, BIM systems have emerged as reliable tools with very high design integration and information flow capabilities. BS1192:2007 was followed by the PAS 1192 suite of standards related to Level 2 BIM data management throughout the building life cycle.

Numerous tools and systems provided by specialist data management technology companies are available, with web-based platforms being the most common. Conject and Asite are examples of such companies, which provide data management solutions ranging from passive data repositories to dynamic BIM-based networks. Their functions are focused principally on data storage, information transfer, file and communication management, and data security. In a CDE, the flow of design and construction information is governed by a standard protocol agreed between the design-construct team members who have access to that information, at all times avoiding one-to-one direct communication, which often leads to information gaps. Figure 9.4 illustrates information and communication exchange in a CDE compared to a traditional flow. In the former, the flow is centrally managed, often between networks, and updates and communications are accessible to all simultaneously. In the latter, however, information flow and communication are heavily dependent on the very high efficiency of team members' internal management processes, a difficult task with the increasing levels of project data and rising expectation for instant communication.

Furthermore, if data-rich project information is uploaded on the CDE, as would be required by BIM Level 2, the flow becomes more dynamic and provides greater opportunities for collaboration. The integrated nature and richness of that information, combined with

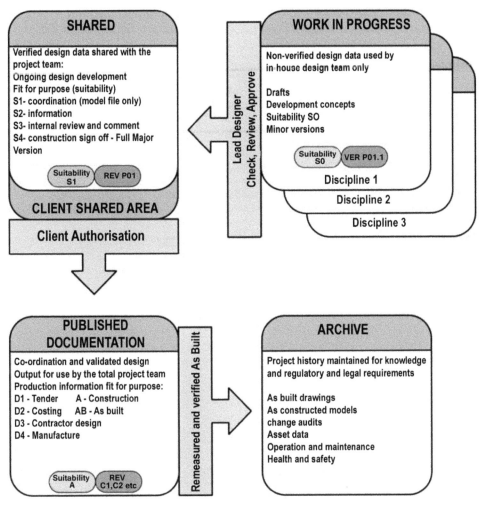

Figure 9.3 BS1192:2007 Collaborative Production of Architectural, Engineering and Construction Information diagram (BSI 2007)

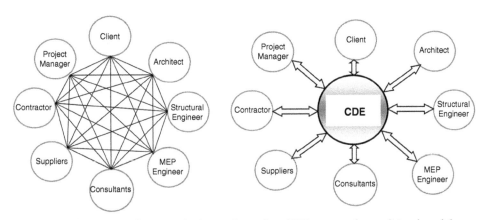

Figure 9.4 Information and communication exchange in a CDE compared to traditional workflow

an efficient data management and control system, would fundamentally change the workflow from a linear, often information-starved one, to a dynamic, data-rich and interactive one.

Bibliography

BIM Excellence (2015) [online]. www.bimexcellence.net/ [accessed 26 March 2015].

BSI (2007) Collaborative Production of Architectural, Engineering and Construction Information Code of Practice, BSi group.

Garrett, G. (2008) 'Heathrow Terminal 5' [online]. http://aecmag.com/case-studies-mainmenu-37/253-heathrow-terminal-5 [accessed 26 March 2015].

HM Government (2011) Government Construction Strategy, Her Majesty's Government, London, [online]. https://www.gov.uk/government/uploads/system/uploads/attachment_data/file/61152/Government-Construction-Strategy_0.pdf [accessed 26 March 2015].

Krygiel, E. and Nies, B. (2008) *Green BIM: Successful Sustainable Design with Building Information Modelling*. USA: John Wiley & Sons.

Latham, M. (1994) *Constructing the Team: The Final Report of the Government/Industry Review of Procurement and Contractual Arrangements in the UK Construction Industry*. London: HMSO.

10 Taking a whole life view

10.1 Developing and applying BIM strategies to meet client and user expectations

The 2013 RIBA Plan of Work begins its stages with the 'Strategic Definition' (Stage 0), which focuses on identifying the client's business case and Strategic Brief and other core requirements for the project. On the other hand, Graham Jones, chair of BIM Steering Group Europe, identifies three conditions for success when building a strategy for BIM at the early strategic stages of a project, and he baptised them the 'why', the 'what' and the 'how' (Jones 2011). For any project using BIM, a strategy needs to be developed or customised to meet the requirements and arrangements of that particular project. Borrowing Graham Jones' terminology, the following will explore the key strategic developments that need to be considered.

The 'why' in the strategy includes a definition of purpose and conducting an appraisal in which targets and objectives are identified and mapped, and a robust business case for the project is well argued. In parallel, a BIM strategy should be developed as part of this overall strategy and fed into the subsequent preparation and brief stage in order to meet the overall client and user short and long-term expectations. Key strategic points to be addressed early include:

- specific (life cycle) information required from the supply chain;
- uptake and adoption of BIM to facilitate compliance with the 2016 mandate;
- consistent handover of digital data in the appropriate format (IFC, Construction Operation Building information exchange (COBie)) and at the appropriate time;
- specification of the required functionality (3D to ND and FM);
- IT platforms and compatibility issues to facilitate information flow and handover; and
- client participation and interaction throughout the process.

The focus of the 'what' is on the comprehensive definition and project deliverables, any relevant precedent experience or examples. Specific definitions of project deliverables will assist in determining accurately the appropriate BIM strategies to be deployed to meet client and project targets; these may vary from the functionalities of CDE, specialist tools or expertise required. In BIM one size does not fit all: each project/client invariably will need a customised solution on a need basis. Size of project, complexity of design and construction expertise of design-construct team members are only examples of variables that may impact on the customisation. A review is needed of previously deployed processes or similar projects undertaken in which tested methods together with any lessons learned could be adapted or reused in accordance with the project structure.

Deployment and implementation, the 'how', is the critical component of any BIM strategy. Regardless of the clarity of purpose and definition of project objectives, if the processes put in place are not well executed, client and user expectations will not be fully met. Prior to implementation the following guidelines should be carefully considered:

- Clarify the overall management structure of the project and how the BIM process fits in, including the definition of roles and responsibilities of all stakeholders.
- Identify any risks likely to a rise throughout the process and put contingencies in place.
- Identify the expertise required, both internal and external, and consider team members' suitability and dynamics.
- Define workflows in relation to project procurement and contractual arrangements.
- Be clear about the project programme targets and how they are to be met.
- Specify collaborative and communication protocols, including a data exchange timetable.
- Define milestones to deliver and meet project and client expectations.
- Define protocols and processes, both internal and external.

Developing and implementing BIM strategies needs careful planning and synchronisation with the overall project programme and client and user expectations. In addition, any strategy needs to be designed within the context of the BIM maturity level of the design-construct team.

10.2 Setting and realising whole life targets for design, construct and use

One of the major developments in the procurement of buildings and infrastructure in recent years has been the breaking down of the barrier that until recently existed between design, construction and operation. The introduction of BIM processes accelerated the integration of design and construction stages, which was then followed by recognition by clients and other stakeholders of the value of integrating the operation stage in the overall asset cycle. Government soft landings standards added greater impetus to the early planning of whole life cycle, as well as asset handover. Looking forward towards an increasingly integrated process of delivery and asset management post construction, setting and realising whole-life targets for each of the above stages becomes critical both for the success of the project and for meeting clients' and users' expectations. BIM as a process is a key enabler in meeting these three stages targets by providing the opportunity for radical improvements in building design, construction and operation through advanced computer-based modelling and data management technologies.

At design stage, targets may include setting standards for carbon footprint, design quality, overall environmental impact, collaboration and co-ordination, project data and information flow, communication protocols, and so forth. To help realise these targets in an efficient and cost-effective way, a BIM system needs to embed these outcomes as part of its processes and protocols. This way, all members of the team are systematically aware of and focused on meeting these built-in targets using the appropriate tools and processes of the BIM system. For instance, specifying a target carbon footprint per square metre in an environmental analysis tool (integrated or as a plug-in with the main BIM software) will automatically throw a fail result until the target value is met.

Interface between design and construction is a good illustration of where the deployment of BIM could have a major impact on a project's quality, cost, time and associated risks. A visualisation of construction sequence, assembly of prefabricated components using manufacturer 3D objects, and pre-construction site layout (including scheduled deployment of plants) are all effective virtual methods of identifying and eliminating risks. Clash detection for design/construction co-ordination is a function of BIM that has proved critical in identifying and eliminating constructability problems traditionally only identified during construction operations often with an impact on delivery time. Using an integrated BIM model, any constructability issues identified late can be rapidly resolved given the instant access to the relevant project information. In addition, flexibility of rescheduling, re-costing and even re-sequencing of building operations will aid in realising targets.

'Hindsight before site' is the ability to prototype the asset life cycle (Jones 2011) at the early planning stage. A BIM model provides early access to useful information that will help better understand the asset, which may be used in a number of ways to meet targets. These may include sales and marketing for a developer, future maintenance/refurbishments cost modelling, and security and surveillance. It is important to keep in mind that the reliability of BIM as an enabler to realising targets throughout the whole life cycle of an asset is dependent on the clarity of overall project objectives and on the level of maturity, as in the latter two criteria will impact on the quality of outcomes, which in turn may help or hinder realising the full potential in meeting targets.

10.3 Engaging clients and building users through simulation and prototyping

At the early design stage of a building, when ideas are still very fluid and very little concrete design material is available, difficulties have always been posed in how to engage clients, particularly when critical decisions need to be made to advance the design process. Beyond technical drawings, architects and engineers historically relied on simple 3D drawings such as perspectives, sketches, etc. and scaled physical models to communicate often design intent and aesthetic. These techniques were developed further with the introduction of CAD, but the principle uses remained largely the same. The introduction of 3D modelling, followed by parametric modelling and the ongoing digitisation of the construction industry, has provided new and innovative methods of communication to designers and clients, which were the preserve of other industries (see Section 12.3). Simulations and prototyping are increasingly replacing older techniques as they are more versatile, command accuracy, communicate better information and are a by-product of the new digital design processes.

Simulations are the most versatile and easy to produce from a BIM model at any stage of the design process. An early model may not contain lots of data but is sufficient to engage with a client in relation to issues relating to siting, urban or rural landscape integration, aesthetics, quality of internal space, planning, etc. Merging visualisations with existing GIS or Google Earth city models, for instance, will provide informative and contextual simulations which clients can easily understand and relate to, and eventually help in making decisions quickly. In the same vein, an advanced model may be used to simulate a client fly-through to ascertain the quality of space and internal finishes, and also provide an interactive capability for the client to change and modify aspects of design, from colour schemes to fittings and materials. Not only are decisions made almost instantly and fed back into a model, but the active engagement of clients and users enhances their experience and increases success rate in meeting their expectations as well as project targets.

Taking a step further from the traditional physical model, a prototype helps to engage clients in visual and tactile ways. It is the nature of people to see and touch objects to appreciate their qualities and functionalities. In a BIM process, data for making a prototype is built in to the 3D model, and little effort is normally required for making a prototype (file formatting and machine set-up). Whether in building or product design, a prototype is used to describe a product more effectively, test the performance of various materials and components, engage clients/potential clients either for feedback or marketing, and test and refine functionality. Their use in building design and construction is wide and may range from a scaled prototype of a whole or part of a building to a detail of a component or assembly or to a particular space. For instance, a scaled prototype of a standard hotel or hospital room will help a client to engage in the accurate and meticulous details of internal fittings and finishes, equipment, services installations, and so on. Further iterations of the prototype can be made quickly once feedback and any redesign has been completed in the 3D model. Similarly, a prototype of an assembly detail may be used by a constructor/manufacturer to test functionality and performance to satisfy standards and eliminate potential buildability problems during construction. Complex design issues, quality of materials and installations, and complex technical details are only examples of the wide uses of prototyping to engage both with clients (to explain intent in a physical and simple way), and with other stakeholders (for common understanding of complex design and construction solutions).

SMEs and micros can exploit the use of these digital techniques when interacting with their clients regardless of how advanced their BIM implementation level is. One might argue that the agility of a small practice in adopting innovative technologies can assist in improving customer service, expanding business opportunities at minimal overheads, especially when using visualisations.

10.4 Embedding soft landings principles into project planning and development

Soft landings is the BSRIA-led process designed to assist the construction industry and its clients deliver better buildings. It is also intended to help solve the performance gap between design intentions and operational outcomes. Embedding soft landings principles into project planning and development occurs at various stages of the project cycle, with the overall intention of safeguarding quality and easing the handover to clients and facilities managers. This section will explore the relevant principles and some of the activities that need to be considered at each of the planning and development stages.

At inception and briefing stages, specific activities related to soft landings need to be added to the client's requirements, developed as part of the project brief and also included in any tender documentation so that budgets are set aside for post-handover care and post-occupancy evaluation. This enshrines these specific activities in the contractual arrangements of the project and makes them mandatory for all relevant stakeholders. Depending on the scope and complexity of the project, specialist consultants or subcontractors may be required to inform on the nature and extent of activities to be included, which will also help in identifying skills and resources required. BIM protocols for data drops handover or transfer of project data onto the client's Computer Aided Facilities Management (CAFM) system are some of the principle activities to be embedded at this early stage.

During design and development stages, precedents should be considered by all project team members to gain insights into tried-and-tested solutions and explore how the future building will be used by end users and facilities managers. Strategies or future activities identified and agreed should be embedded into each team member's relevant design, which later will be incorporated in the BIM model as appropriate. Some of the principles and related activities to be agreed and embedded at these stages include the energy strategy and targets, in-use energy monitoring regime, commissioning strategy and timetable, and usability and maintainability of installations. These principles need to be reviewed and updated as the design develops for feasibility and buildability purposes.

A phased handover of the completed building should be planned and embedded in the overall soft landings strategy. This will mostly impact contractors' and subcontractors' schedules, which should be managed and updated in accordance with the handover strategy. A phased handover avoids overwhelming, particularly in complex projects, operators and FM managers and enables them to spend more time understanding interfaces and systems before occupation. Activities to plan for may include the practicalities of project data handover, BMS set-up to initial client requirements, energy monitoring software and any initial training on management or monitoring software. BIM data from design-construct teams need to be as-built updated, checked and handed over in the specified format.

Post-handover, a period of six to eight weeks is recommended to be scheduled by project team members, during which they need to deal with any emerging issues related to system operations, software malfunctions, etc. It is also critical at this stage that all aspects and systems of the building are checked against the client's soft landings requirements and meet their expectations. Longer-term, Post-Occupancy Evaluation (POE) should also be embedded and may include activities related to long-term energy performance, reliability of installed software systems, and any discrepancies on project data. As recommended by soft landings guidelines, systematic POEs are to be conducted no sooner than 12 months post-handover, then repeated at 12-month intervals, culminating in a final project review at month 36.

Bibliography

BSRIA, Soft Landings, [online]. https://www.bsria.co.uk/services/design/soft-landings/free-guidance/ [accessed 15 April 2015].

Jones, G. (2011) Building a Strategy for BIM, [online]. http://cic.org.uk/admin/resources/dl-cic-bim.pdf [accessed 14 April 2015].

Prototyping, [online]. www.usability.gov/how-to-and-tools/methods/prototyping.html [accessed 5 April 2015].

RIBA Plan of Work (2013) [online]. www.architecture.com/files/ribaprofessionalservices/practice/ribaplanofwork2013overview.pdf [accessed 15 April 2015].

11   Managing information flow

11.1 The value of smooth information flow in meeting objectives

Delivering construction projects to meet the triangle of quality, cost and time and to the satisfaction of the set objectives is heavily dependent on the flow of information through the various design and construction stages of a project. Built environment projects involve huge volumes of information that have to be transferred between numerous different project stakeholders throughout the building life cycle. Further, much of that information is graphical in format. Over the years, built environment professionals have evolved ways of managing that volume of information by employing various tools and techniques, such as a project client brief, which still remains valid in a fast-changing industry landscape peppered with new working practices, new technologies and evolving contractual and legislative frameworks, amongst others, that are central to deliver to the drawing registers. However, being able to identify the current information you require, retrieve it and manage its flow efficiently is still far from easy in most projects.

Clients tend to recognise the value of that flow in shaping collaborative working and are increasingly insisting that their design-construct teams demonstrate efficient collaborative capabilities when making their choices. They also recognise that an efficient project collaboration tool will assist in the development process, including the effective management of information flow. Furthermore, the complexity of construction projects, the associated large quantities of data to be processed and the ever-increasing statutory and technical requirements render the smooth flow and management of project information a key component to successfully meeting objectives.

As illustrated in Figure 11.1, traditional information flow is fundamentally linear and follows a phased sequence of design and construction activities predominantly based on the RIBA Plan of Work model. Project information tends to be contained within each phase, such as design or procurement, until such a time where it flows to the next phase and remains largely invisible to other design-construct team members, particularly members with minor contributions such as a small ME subcontractor. Collaboration is very limited by virtue of the phased nature of the flow and incompleteness of information between phases. In a collaborative BIM working environment, however, the flow of information is continuous and up-to-date, and access to all stakeholders is automatic within project protocols. The bi-directional flow between each contributing member and the model (single or federated) simplifies the flow process and adds transparency. Simultaneous working and live interactions of team members via a BIM model in a Common Data Environment enables the highest levels of collaboration and smooth data exchange.

The value of smooth information flow and the collaboration potential it generates resides not only in delivering project objectives but also in creating long-term working relationships and partnerships and meeting wider and global goals such as CO_2 reduction.

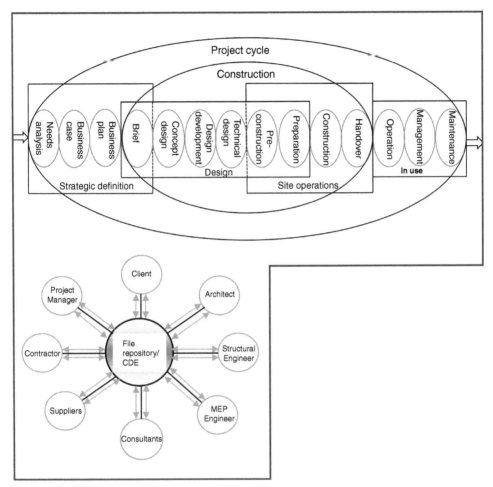

Figure 11.1 BIM (multidirectional - *bottom*) compared to traditional (linear - *top*) information flows

11.2 Setting up and managing CDEs

The Common Data Environment (CDE) concept developed at the emergence of Software as a Service (SaaS) collaboration platforms in the early 2000s and, as mentioned in Section 9.4 above, their use intensified in response to British Standard BS1192:2007. This was later strengthened by the UK's BIM push strategy and the publication of PAS 1192-2, specification for information management for the capital/delivery phase of construction projects using Building Information Modelling. This shift to BIM has helped concentrate minds beyond systems to manage electronic documents. Instead, discussions are more likely to be focused on how to manage the 'I' in BIM – the information, or, more specifically, the data at the heart of information exchanges. And instead of managing files, the collaborating design-construct team members will increasingly be managing the core data contained in the model to, initially, comply with Level 2 BIM requirements. This will involve multidisciplinary models combined into a 'federated model'. There is an implied expectation that the industry will gradually migrate towards iBIM or BIM Level 3, where a single shared model containing all project information is created and progressively improved throughout the delivery cycle to handover stage, when the BIM-generated data is passed on to the owner/operator of the built asset for future management purposes.

Setting a CDE is normally initiated by a client or a contractor, often in a design and build contract. Large clients such as food chain stores or government agencies and local authorities use CDEs to quality manage their wide building asset portfolio and maintain a level of control. It is a bespoke service provided by IT companies specialising in project information management. An assessment of the level of service to be provided is made, followed by agreeing terms and conditions of service between the initiator (often client or contractor) and the information management system provider. A package catering for the level of service to be provided is set up by the latter, normally in a short period of time (ranging from one to four weeks depending on the complexity of service). The service provided may range from a simple repository and file management system (as required by BS1192:2007) to a fully fledged collaborative BIM model environment. The flow of information from and to that model, as well as all interactions between design-construct team members, is managed by the CDE service provider. However, the content of the model or data files, quality of information, updates or merger of team models for clash detection and resolution remain the responsibility of the BIM co-ordinator, a member of the project lead. As an example, Conject describes its 'Collaboration with control' service functionality as follows:

> A functionality rich Common Data Environment (CDE) enables PAS 1192 & BS1192 compliant working practices acting as a secure, central fully auditable information repository from which documents, models and information are shared, accessed and managed by authorised users. CONJECT's CDE continues to deliver this trusted capability with BIM, as it has done for over 14 years with 2D drawings.
>
> (Conject 2015)

Key to the functionality of a CDE is using a collaboration platform provided by the CDE provider to facilitate access and manipulation of project data and capable of handling multiple file/model formats including IFC format. The latter is increasingly used to overcome issues of compatibility, which are a key component of any BIM or CDE protocol. Access and use of software platforms by design-construct companies has changed since the late 1990s as a result of the rapid expansion of the Internet. Instead of a fixed number of licensed copies of a software installed and managed on individual machines and managed by in-house IT technical staff, these organisations may rent software from software providers who also provide technical support. This Software as a Service (SaaS) is delivered via the Internet to where it is needed on both fixed and mobile devices. This model is known as cloud computing, where software and company data is stored in a network of remote servers rather than a fixed location. The benefits of SaaS include:

- it is easy and cost-effective to roll out across a diverse supply chain;
- there is no limit on the number of users;
- it typically can be accessed through a range of internet browsers or mobile devices;
- there is no need for IT departments to manage or maintain the software;
- no plug-ins, additional software or firewall configurations are required;
- it can be used by any member of the team, anywhere;
- data security and recovery are part of the service; and
- software is updated regularly.

CDE and SaaS are increasingly used by operators in the industry, including SMEs, who could find added value in introducing/adopting these innovative practices while developing a new business model and workflows leading to Level 2 BIM.

11.3 The BIM/information manager role: explicit or embedded

Subsequent to the migration from drawing board to 2D CAD that took place in the 1970s and 1980s, management of design and construction information using new software packages and associated filing systems became the preserve of a new professional with specialist technical and managerial IT skills. The role became known as the CAD Manager. The skills required of the new role could be characterised as predominantly technical and largely related to the software application being used, e.g. functionality, layering of information, etc. The design and construction knowledge, as well as the methods of graphical representation, remained largely unaffected apart from the medium, i.e. board to screen drafting. It took the industry 20 years to complete that migration, a figure often quoted in the industry literature.

In contrast, the current migration to collaborative BIM is more complex on both the technical and knowledge counts. The technical one involves new skills related to building modelling instead of drawing and, more crucially, embedding multifaceted parametric data in that model. On the knowledge count, an understanding of how buildings function, perform and are assembled is required beyond the traditional junior CAD technician level. Modelling involves an advanced level of decision-making, specification and collaborative interaction early on and throughout the design/construction process. In addition, the level of information and metadata in a BIM model is extensive and requires specific skills not only in how it is to be generated but also, most critically, in how it is managed effectively. Using filters, merging models in a federated or central model environment, clash detection/resolution and running performance simulations are only examples of the skills needed.

Drawing parallels with the CAD Manager, a new role has emerged in recent years and is often referred to as the BIM or Information Manager (occasionally also labelled Design Manager). The visibility of this role within the industry may be explicit and well defined within the day-to-day operations of a built environment organisation, often a large Architecture Engineering Construction (AEC) company. Or it may be an embedded role performed individually or sometimes collectively by members of a design-construct team, often in a small SME where roles tend to overlap due to organisation size and versatility of personnel. It is important, however, to distinguish between the emerging role of Information Manager and BIM Co-ordinator as defined in the CIC Protocol (CIC 2013a): 'The Information Manager has no design related duties. Clash detection and model co-ordination activities associated with a "BIM Co-ordinator" remain the responsibility of the design lead.'

In addition, the protocol provides for the appointment of an Information Manager. Part of the project contractual arrangement, it requires the Employer to appoint a party to undertake the Information Management role. This is expected to form part of a wider set of duties under an existing appointment and is likely to be performed either by the Design Lead or the Project Lead, which could be a consultant or contractor at different stages of the project. There is also the provision that, in some circumstances, the Employer may appoint a stand-alone Information Manager. The scope of service of the role is outlined in a separate CIC document entitled 'Outline Scope of Services for the Role of Information Management' (CIC 2013b). In brief, the scope is organised under three areas of responsibility:

- Common Data Environment Management includes establishing a CDE, maintaining the model, managing common data and validating model information.
- Collaborative working, information exchange and project team management includes supporting the implementation of the project BIM protocol, supporting collaborative working and information exchange processes, and complying with team management procedures and processes.
- Project Information Management includes initiating, agreeing and implementing the Project Information Plan and Asset Information Plan.

Explicit or embedded, the Information Manager's role in a BIM collaborative environment is critical to the set-up of the CDE, the flow of information and compliance with agreed protocols and procedures.

11.4 Integrating workflows to realise whole life targets and outcomes

Built environment professionals are very familiar with file formats such as DWG, DXF and PDF, which have for many years been used either for sharing or accessing design and construction data files across software platforms. These formats are still commonly used in practices where 2D drawings are still prevalent including data repositories. The recent shift to using parametric modelling software and associated BIM workflows has been a game changer in the way project information is shared and accessed in a collaborative working environment. In the latter, smooth workflows are essential to collaboration and any barriers will not only hamper teamworking but also limit the benefits that may be derived from that collaboration. Realising whole life targets and outcomes of an asset requires careful integration of these workflows throughout all stages and processes, from inception to decommissioning.

A number of issues that may affect the integration of workflows are explored here. Interoperability of project data across platforms, that is the ability to access and utilise BIM data model files authored by different software platforms, was until a few years ago an obstacle for a variety of reasons, ranging from purely commercial to technical. The International Foundation Classes (IFC) format, a Building smart standard, is increasingly becoming the accepted standard for open BIM, allowing for full interoperability between BIM authoring platforms. Wider adoption of open standards by the range of software platforms used in the construction industry is vital for smooth workflows and critical to the success of the Government's strategy for BIM adoption. Linked to data standards in the Government BIM strategy is the quality and reliability of asset data to be made available to clients, known as data drops in a COBie standard. A complete COBie should be expected by clients at the time of handover, but earlier interim drops at key stages of the delivery cycle are used to monitor the business case for the facility and to help plan for taking ownership. Once received by the client, the COBie information can either be kept as delivered or held in ordinary databases, or it can be loaded into existing Facility Management and Operations applications, either automatically or using simple copy-and-pasting (COBie 2012). This full asset data handover marks the start of the facility management cycle of the workflow. Also essential to the integration of workflows is for the design-construct team to agree, as early as feasibly possible, all protocols and standards. These will include, amongst others, those related to information management, data exchange, data format standard, etc. CIC protocol and the mandated information manager role will, when fully applied, assist in the integration of workflows and the overall management of project information in a collaborative BIM environment.

In conclusion, each building asset will be expected to realise well-defined targets and outcomes at different stages of the life cycle and to different stakeholders. Adopting, early on in the process, an integrated approach to the set-up, implementation and management of workflows will assist in meeting the set targets and outcomes.

References

CIC (2013a) Building Information Model (BIM) Protocol, [online]. www.cic.org.uk/download. php?f=the-bim-protocol.pdf [accessed 31 March 2015].

CIC (2013b) [online]. www.bimtaskgroup.org/wp-content/uploads/2013/02/Outline-Scope-of-Services-for-the-Role-of-Information-Managment.pdf [accessed 31 March 2015].

COBie (2012) [online]. www.bimtaskgroup.org/index.php?s=COBie [accessed 31 March 2015].

Conject (2015) [online]. http://conjectblog.co.uk/2014/09/what-is-the-common-data-environment/ [accessed 1 April 2015].

12 BIM futures

12.1 Established and emerging environments and techniques for data management

Common Data Environments (CDEs) may be thought of as the established method of data storage and management systems for construction projects, bar small ones. As highlighted in Sections 9.4 and 11.2, BS1192 and PAS 1192-2 have been major contributors as push factors. Use of CDE is far from universal in practice; SMEs, micros and even larger organisations will continue using their existing data storage and management infrastructure for small-size projects, particularly where cost overheads are prohibitive. However, they may well use other forms of data repositories, Dropbox or Google cloud, for sharing files. This hybrid use may continue over the transition period to BIM beyond the 2016 deadline. The diagram in Figure 12.1 illustrates the functionality of a CDE in terms of data flow and interconnectivity of design–construct team members. The biggest value in using CDE, especially to PAS 1192 standard, resides in the collaborative potential through the workflows from inception to commissioning.

High speed internet access combined with the increasing ease of accessing the cloud, where the potential for inexpensive data storage, may emerge as the next working environments for those in transit to full BIM. Combined with Software as a Service, small practices may find opportunities in selecting services on a business-need basis from the range of services provided by technology and software providers. The open subscription currently operated by major software companies makes it easier to access the range of BIM packages, plug-ins and small applications, and if exploited correctly, could add to the agility of small practice not only in making a rapid transition to BIM but also in exploring new business opportunities in a fast evolving construction market. Apps on mobile devices such as mobile phones and tablets are also emerging as nimble techniques of data management capable of being remotely synchronised with larger networks through mobile Wi-Fi. For instance, 123D Catch is a free app from Autodesk that lets the user create 3D scans of virtually any object from a digital photograph, as illustrated in Figure 12.2. Once the 3D scan is complete, it can be scaled for insertion in a virtual environment, 3D printed or used in an augmented reality setting. All may be done from a single mobile device and streamed via mobile Wi-Fi to any remote location.

12.2 Big data, the Internet of Things and interactive technologies

In our day-to-day life we are bombarded with large amounts of data that most of us are unable to process or make sense of immediately. The widening spread of the Internet and the

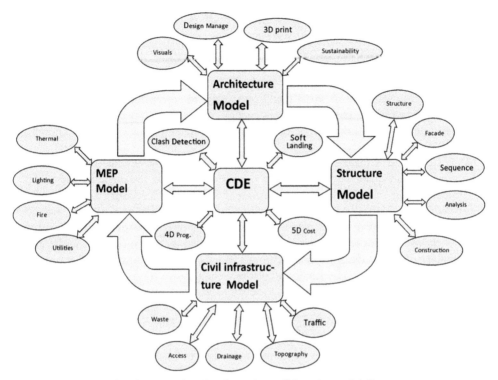

Figure 12.1 Functionality diagram of a CDE for project collaboration and delivery

Figure 12.2 3D model generated using 123D Catch app (final year work of architectural technology students, Robert Gordon University 2014)

proliferation of internet-enabled fixed and portable devices are rapidly changing not only daily life but also how we interact with each other both on a personal and professional level. Take a very simple example: up to 20 years ago taking photographs whether on holiday or on a building site involved the logistics of buying a film, taking the photos, processing them with specialist equipment and then finding a physical space to store them. Now all this is done instantly on a pocket digital camera or a mobile phone and in addition the photographs can be shared instantly via email or social media and safely stored on the cloud. In the professional world, be it finance, transport or construction, the scale of data collected, stored and manipulated on large networks is staggeringly high and increasing, now commonly known as big data (BD).

BD is a broad term for data sets so large or complex that conventional data processing applications are inadequate. The challenges in processing these data sets include analysis, capture, search, sharing, storage, transfer, visualisation and information security. The term often refers to advanced methods to extract value from data or to the use of predictive analytics, and seldom to a particular size of data set. In the context of construction, and particularly BIM models loaded with large parametric data (often called metadata), analysis can be used, for example, to identify clashes when merging various consultants' models. Other applications include code checking, buildability simulations, comparative cost analysis, and predicting alternative solutions and trends in scenario simulations, amongst others.

Relational database management systems, desktop statistics and visualisation packages increasingly have difficulties in handling big data. Processing requires parallel software running on a large number of servers through linked networks. In recent years, relational databases have been challenged by object databases, now commonly used in BIM software. To put it in context, what constitutes big data varies depending on the capabilities of the users and their tools, and increasing capabilities make big data a moving target. Therefore, what is considered to be 'big' in a particular year will become ordinary in later years. In an increasingly data-hungry BIM collaborative work environment, for SMEs and micros the prospect of facing hundreds of gigabytes of data for the first time may trigger a need to reconsider data management options. For others, larger BE organisations, it may take tens or hundreds of terabytes before data size becomes a significant consideration.

The Internet has become an enabler of how we process and exchange data and is pervasive in all aspects of life from banking to retail and particularly embedded devices through linked networks. The latter is termed as the Internet of Things (IoT). IoT is defined as the network of physical objects or 'things' embedded with electronics, software, sensors and connectivity to enable them to achieve greater value and service by exchanging data with the manufacturer, operator and other connected devices. Each thing is uniquely identifiable through its embedded computing system but is able to interoperate within the existing internet infrastructure. Applications of IoT, in the construction sector may include sensors for monitoring HVAC (Heating, Ventilating and Air Conditioning) systems for energy consumption or embedded devices in building components to collect performance data or in a simple house boiler. The expectation is that IoT will offer advanced connectivity of devices, systems and services beyond machine-to-machine communications and will cover a variety of protocols, domains and applications. Large amounts of data from diverse locations are also expected to be generated via IoT, which will be aggregated very quickly, thereby increasing the need to better index, store and process this data.

Embedded devices may stream data directly through Wi-Fi to networks and data storage locations automatically, but accessing them directly through interactive technologies offers another advantage. For instance, an embedded device in a prefabricated concrete facade panel will collect and stream data on water penetration, wind load levels, temperature swings, etc. The data may be fed automatically to a network or could be picked up locally via a portable reader device, such as a mobile phone or a tablet, using a bespoke application. These interactive technologies are commonly used in other industries. Examples are transponders on aircrafts, boats and even automobile tracking devices linked through satellite and GPS systems. Big data, the IoT and interactivity are all new technologies and innovative practices expected to be part of BIM futures in a future digital Britain.

12.3 Prototyping and digital fabrication of components, assemblies and whole buildings

The design and construction of a building has for long been seen as a one-off enterprise in which a blend of art, human experience and craft produces a unique structure. That perception altered with the modern movement as expressed by Le Corbusier in 1923: 'a house is a machine for living in' (Le Corbusier 2008). Industrialisation of construction followed throughout the twentieth and twenty-first centuries, but not to the extent of other industries such as the automotive and aviation industries. In the manufacturing industries, practices such as prototyping, digital fabrication and robotic assembly of components are common practice and have been for decades. Their use in the construction industry has been slow but has accelerated recently with the advent of digitisation and BIM.

In essence, a prototype is a draft version of a product that allows you to explore your ideas and show the intention behind a feature or the overall design concept to users before investing time and money into development. Model making is a well-established practice in architectural design, a type of prototype, but the focus tends to be on the representation of space and aesthetic rather than a real prototype of the building to test and explore performance and fitness for purpose. The real obstacle, one could argue, has been the absence of technologies and processes similar to those in other sectors. It is much cheaper to change a building component early in the development process than to make changes after you have completed the whole building. Prototyping allows the gathering of feedback from testing and experimentation for performance and usability. Milling, laser cutting, 3D printing and CNC tooling are only some of the digital and non-digital processes that are increasingly being used in building component manufacturing and, most importantly, new generations of BE graduates are being trained in these new technologies.

Digital fabrication is a type of manufacturing process where the machine used is controlled by a computer. The most common forms of digital fabrication are:

- CNC machines, where shapes are cut out of material sheets;
- 3D printing, where objects are built up out of layers of metal or plastic; and
- laser cutting, where materials are cut, engraved or burnt (or melted, in the case of metals) by a laser beam.

There is a huge range of digital fabrication techniques. The important aspect that unifies them is that the machines can reliably be programmed to make consistent products from digital designs. Take for instance 3D printing, an emerging digital technology in the construction sector. After having built a BIM model of the building a designer can at a click of a button 3D print a scaled prototype of the whole or any part of that building. Scale and level of detail may be chosen in relation to the intended purpose, and a junction detail may be explored for buildability, assembly of components, etc. 3D printing parts or a whole building is an emerging

Figure 12.3 Experimental 3D printed house from waste material (after De Zeen Magazine 2014)

technology, currently experimenting with a variety of materials and assembly systems, and when established would represent an example of full transition to digital fabrication in construction. This way, the processes and associated efficiencies enjoyed by other industries may start making their mark in construction.

12.4 Augmented reality and virtual environments

Video games have been entertaining us for a number of decades, ever since the very first games were introduced to arcades in the early 1970s. Since then, computer graphics have advanced and become much more sophisticated. Graphics in the gaming industry are pushing the barriers of photo realism. That is pulling graphics out of a computer display or television and integrating them into real world environments. This new technology, known as augmented reality (AR), blurs the line between what is real and what is computer-generated, by enhancing what we see, hear and feel. Compared to virtual reality (VR), which creates immersive, computer-generated environments, and the real world we live in, AR is closer to the real world. It adds graphics, sounds and haptic feedback to the natural world as it exists. Augmented reality is changing the way its users see the world. Whether it is Google glasses or Apple smart devices, these technologies will make their way into the design and construction of buildings and will add a whole other dimension to how built environment professionals and users interact with the buildings.

On the other hand, developments in virtual environments (VEs) are helping to create interactive and immersive experiences for all users. VEs may be deployed for any use – the popular Minecraft game is a vivid example; simulators for pilots or military training are other examples. VEs are generated by software that implements, manages and controls multiple virtual environment instances. In the Architecture Engineering and Construction (AEC) sector,

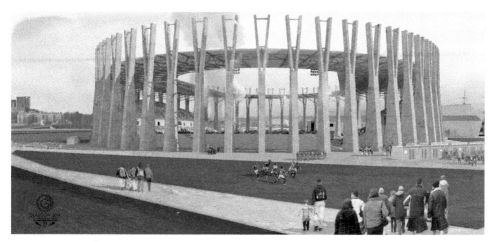

Figure 12.4 Virtual environment simulation around a 3D model design of a football stadium (final year work of architectural technology students, Robert Gordon University 2014)

virtual environments of buildings are increasingly created by architects and engineers using BIM authoring software for various purposes. These may include building/site visualisations for client and planning authority approval or construction assembly sequences or purely for sales and marketing purposes. Figure 12.4 illustrates a virtual urban environment simulation around a 3D model design of a football stadium.

References

De Zeen Magazine (2014) 'Chinese Company 3D Prints 10 Buildings in a Day Using Construction Waste', [online]. Available at: www.dezeen.com/2014/04/24/chinese-company-3d-prints-buildings-construction-waste/ [accessed 3 April 2015].

Le Corbusier (2008) 'Towards a New Architecture', BN Publishing [online]. Available at: http://www.barnesandnoble.com/w/towards-a-new-architecture-le-corbusier/1006079234?ean=9789650060367 [accessed 3 April 2015].

Part IV

Setting up a BIM project

13 SMEs, micro–SMEs and BIM

13.1 Organisational structures and business models

Small and medium-sized enterprises (SMEs) and micro-SMEs play a central role in the European economy. Although the 'average' European business employs no more than six people, SMEs and micros are a major source of entrepreneurial skills, innovation and employment. In the enlarged European Union with 28 member countries, some 23 million SMEs provide around 75 million jobs and represent 99 per cent of all firms. However, as EU studies have identified, SMEs and micros may be faced with significant market challenges, particularly in obtaining capital or credit, in the early start-up period. As limited resources can inhibit access to new technologies and innovation, support for SMEs is one of the European Commission's priorities for economic growth, job creation and economic and social cohesion.

The EU defined SMEs under three qualifying categories according to the number of employees, and annual balance sheet total or annual turnover (Anon 2005):

- micro: fewer than 10 employees; annual turnover or annual balance sheet total equal to or less than €2 million;
- small: fewer than 50 employees; annual turnover or annual balance sheet total equal to or less than €10 million; and
- medium-sized: fewer than 250 employees; annual turnover equal to or less than €50 million or annual balance sheet total equal to or less than €43 million.

BIM4SME was the working group given the mission by the Government BIM Task Group of taking the Level 2 BIM message to the UK construction SME and micro-SME community of around 250,000 firms. BIM4SME (officially launched in April 2013) is a pan-industry volunteer team of professionals drawn from diverse supply chain backgrounds, including client, design, construction, supplier and industry organisations. Since the launch, BIM4SME has actively promoted the interests of small and medium-sized firms through diverse interactions with a wide range of construction interests. These have ranged from dialogue with strategic industry groups, the CIC BIM Hub network and professional bodies, to local initiatives such as running regional BIM events and practitioner BIM clinics.

Chaired by Tim Platts, the BIM4SME team is characterised by a vision and collective passion for driving digitisation within the industry forward. As a paradigm for pan-discipline collaboration and collective purpose, the forward travel and momentum achieved by BIM4SME has been significant in the two years between 2013 and 2015. The focus for that activity has been in championing SME interests across UK construction supply chain organisations. With such a large, diverse and diffuse SME presence, BIM4SME's biggest challenge has been reaching its target audience en masse. In the absence of metrics to benchmark, test and measure industry progress with BIM uptake, it is difficult to assess how effective the team's efforts have been in galvanising change. Certainly, if Government's aspiration for a digitised sector is to materialise, UK construction needs to embody change. The process of stimulating the technological and cultural shifts begins with having SMEs and micro companies on board.

As an enabling organisation, BIM4SME has made considerable efforts to get to grips with that challenge, engage with the UK SME community and promote dialogue. As landscape architect Henry Fenby-Taylor of Colour Urban Design put it:

> having met with the BIM4SME group recently, I am inspired. I have been supporting the implementation of BIM at Colour Urban Design and through this I have made contact with a number of thought leaders. Most of these are part of large organisations with significant resources behind them. Sometimes you could be forgiven for thinking BIM exists just for them.
>
> (Fenby-Taylor 2015)

SMEs and micros tend to be characterised by relatively flat organisational structures with little or no vertical management hierarchies between salaried employees and executive decision-makers. Jason Fried, co-founder of Chicago-based digital start-up 37signals (later recast as Basecamp) reinforced the point:

> Besides being small, 37signals has always been a flat organization. In fact, flatness is one of our core values. We have eight programmers, but we don't have a chief technical officer. We have five designers, but no creative director … Even as we've grown, we've remained a lean organization. We do not have room for people who don't do the actual work … One thing we've found is that groups that manage themselves are often better off than groups that are managed by a single person. So when groups do require structure, we get them to manage themselves.
>
> (Fried 2011)

The ability to adapt to change quickly and efficiently is said to characterise SMEs and micro companies. Data trawled from a 2008/9 global survey conducted by The Economist Intelligence Unit cited MIT research which suggested that agile firms grow revenue 37 per cent faster and generate profits 30 per cent higher than non-agile organisations. Caroline Stockmann, formerly of Unilever, now Chief Financial Officer of the British Council, offered three lessons for improving organisational agility in tough times: Lesson #1, devote as much rigour to management before a crisis as during it; Lesson #2, don't let communication and teamwork lapse; Lesson #3, be wary of the status quo.

Because digital technologies now underpin nearly every contemporary business activity, they have the potential to be applied to the management of business-critical data independent of organisational size and structure. Technology's role as a change agent can also enable companies to adopt best-in-class knowledge-sharing processes. Somehow, the idea that the usefulness of information transfer between BIM models depends on the ability of different file types to be able to interact seems archaic in a digital age. Two-thirds of respondents to the 2008/9 Economist Intelligence Unit survey wanted to see technologies which pulled data from multiple applications used for product and service innovation. That pointer alone puts down a significant marker for BIM as a facilitator of process. To some extent it also challenges the BIM level transitions paradigm embedded in the UK Government BIM Task Group's strategic thinking.

Many UK design, construction and building management/maintenance SMEs and micros will have some association with one or more professional bodies and trade organisations. These affinities represent extended organisational networks. A multidisciplinary practice might have employees drawn from a range of professional bodies. Visceral forces may influence associations and polarisations between professional institutions, particularly in times of fiscal restraint when workloads dip and financial returns are under pressure. Technology generally and BIM specifically would be a case in point where new and emerging digital tools may challenge established roles, responsibilities and workplace practice. Professional bodies tend to have committee-driven complex organisational structures. These may be inherently less agile than smaller, flatter hierarchies and limit the ability of the built environment institutions in adapting to change quickly and efficiently in supporting their constituencies and members.

Stan Lester, in his 2009 study of UK professional bodies, highlighted various socio-economic, technological and educational factors acting on professions and their umbrella institutions. As drivers for change, it was noted these may not be evenly balanced (Lester 2007). In 2007, the American Institute of Architects (AIA) introduced Integrated Project Delivery (IPD) to the US construction market as an holistic whole life paradigm for delivering construction projects by integrating people, systems and practices from the start of the design phase. BIM was represented as a key technological enabler. It is a commendable initiative, but templates for professional practice must be underpinned by developing and applying appropriate pedagogies supported by both educators and industry. To what extent evolving and best practice for BIM is filtering down from professional bodies to their constituent SMEs and micros in the UK remains an open question in the meantime. Clearly, in terms of Lester's model, education may represent a key factor in the drive towards more diffuse and inclusive implementation of BIM across built environment professions.

In 2010, the West Yorkshire Lifelong Learning Network recorded over 40 organisations sharing common interests in built environment education and training in the UK (WYLLN 2010). The Construction Industry Council (CIC) is UK construction's umbrella organisation representing professional bodies, research organisations and specialist business associations. Drawing on the experiences of one CIC member, a 2014 trawl of Chartered Institute of Architectural Technologists's (CIAT) accredited undergraduate degrees in Architectural Technology revealed that while some universities have already introduced BIM into the curriculum as a course-specific subject, others are using BIM to leverage broader and deeper interactions between built environment disciplines.

One academic course leader noted that for a final year multidisciplinary project involving 250 students, the overriding objective was 'to develop the course placing equal importance on students experiencing both the people aspects of collaboration as well as the technical skills, with no bias towards either'. Does one university's actions represent a wake-up call for others? Certainly, any advocacy for BIM education embodying interactions which bridge perceived professional boundaries seems positive, and there have been precursors. Strathclyde University's pan-discipline Building Design Engineering (BDE) undergraduate degree (accredited across a range of built environment professional disciplines) ran from the early 1990s until around 2007. Information technology was a core study subject.

Perhaps not surprisingly, early BIM adopters and large projects have been widely publicised in the media. In that context, some aspects of BIM practice, such as the COBie protocol, have been prototyped, tested and shared across the industry. But to what extent BIM processes are agnostic and scale down to fit SME and micro business models deserves equal coverage, rigorous scrutiny and evaluation, as does the extent to which BIM challenges the resources, less of bigger players and more of the smaller partners within supply chain structures.

The Australian Glenn Murcutt achieved stellar status as a sole practitioner, winning a string of international awards and, as American architect and teacher Francois Levy demonstrated convincingly in his book *BIM in Small-scale Sustainable Design* (2012), doing BIM does not necessarily involve a portfolio of large projects and complex organisational structures. Similarly, as director of a firm sitting in the 'small' category of SMEs (fewer than 50 employees), Spencer Fereday of SME structural engineers Mann Williams described how his company philosophy shifted from BIM seeming irrelevant, to engagement, then managed the transition from lonely BIM towards design for manufacturing. In engaging with that journey, Fereday noted that Mann Williams substituted 'own practice' templates for 'Government' BIM paradigms early on in the process; BIM (including the application of standards) has to fit within firms' business models.

Incorporating BIM into SME or micro business models is not predicated on increased office space or huge capital investment to support new equipment and infrastructure. In fact, in terms of office overheads the reverse could apply. Ubiquitous tools such as Skype and GoToMeeting can serve online working from any location offering an available Wi-Fi connection. File sharing freeware like Dropbox can be upgraded to more sophisticated software as business needs require. Over the past two decades, the demand for wireless service has grown at an extraordinary pace. Wi-Fi hotspots are offered in an increasingly wide range of public locations (shops, restaurants, airports and even ferries), and technologies under development are intended to boost the speed of wireless local area networks towards 6.75 Gbit/s over distances less than 10 metres.

SMEs and micro businesses are already supported by laptops, PDAs and mobile phones. Typically, these devices can run applications like multimedia imaging for client presentations, remote conferencing, design/construction team meetings and video streamed CPD webinars. At this point in time, the most significant constraint limiting mobility may be maintaining a viable Wi-Fi signal out of range of hotspots. Dependency on mobile phone service providers can produce variable results in different geographical locations.

Homeworking is no longer viewed as the domain of immature business models. The information technology infrastructure is already in place to enable small organisations to establish and maintain a competitive edge with BIM. Rigorous self-management, smart teamworking and quick response to new work practices, which could enhance quality of output and profitability, and which are Caroline Stockmann's trilogy of pointers for maintaining competitiveness in tough times, are all applicable. SMEs and micros have the agility to change direction quickly and embrace BIM practice and workflows.

13.2 Resourcing and future planning for BIM uptake and deployment

Spencer Fereday described the learning curve as his organisation migrated from lonely BIM to Design for Manufacture and Assembly (DfMA) from late 2011 onwards. 'Although in the depths of recession and competing against some suicidal fee bids we looked strategically at BIM' (Fereday and Potter 2013). One of the key challenges was to shift focus from the Government's BIM templates to a practice-centric paradigm, initially by questioning whether BIM could provide a fee percentage for the business and return profitability to pre-recession levels.

For Mann Williams, with a 30-strong complement of consulting civil and structural engineers divided between offices in Bath and Cardiff, the implementation and use of BIM presented significant financial and technical challenges. Based on an analysis of three projects executed between 2011 and 2013, Fereday concluded that the initial investment in BIM had improved profitability and boosted staff confidence about the future of the construction industry. He suggested that a key issue for SMEs (and perhaps the industry in general) was to ensure the costs are proportionate to the level of BIM data required by clients and contractors.

David Miller Architects (DMA) is a small practice with a varied workload which operates from an office in central London. David Miller has been an enthusiastic BIM advocate for a number of industry platforms. These have included chair of the Construction Industry Council (CIC), London BIM Hub and membership of BIM4SME, a pan-discipline working group of the Government BIM Task Group. Rapid growth of the practice (from four to fourteen staff) was driven by incremental strategy for BIM implementation. DMA argued that using BIM throughout the entire pre-contract process has ensured fully co-ordinated, accurate and consistent data output. Integrated BIM models have combined architectural with structural and services data. In some instances (like Mann Williams) that joined-up process has been extended to link with off-site construction techniques. For example, it was used at the flat-packed Mayfield Secondary School for the London Borough of Redbridge, where intensive collaboration at the design stage facilitated achieving the client's 18-month programme target.

David Miller noted that purchasing BIM authoring of software was only the beginning of an evolutionary journey (Figure 13.1) and reflected that his office had become steadily more sophisticated over the last four years, with an intensive push towards collaborative BIM over the last two:

> It would have been a considerable challenge for us to go straight to level two BIM without a few practice runs. At DMA we were lucky to be able to cut our teeth on some smaller projects where if it had gone wrong we could have easily reversed out into 2D CAD. We started with some initial training and went straight into a small £1m residential building, using Revit simply to produce the drawings, with no collaboration.
>
> (Miller 2012)

Writing for the NBS in 2012, David Miller noted:

> We have come to realise that small organisations like us have got it easy when it comes to change-management. While the cost is often seen as the barrier to entry (it is the same for larger practices, only scaled), it's actually the process change that is the real challenge. Here smaller practices have the upper hand; we don't need to convince the board or investors and it's easy for us to overcome internal resistance from staff with a vested interest in doing things the way they always have. So whether you approach BIM through ROI (return on investment) calculations or you act on instinct and experience, a small practice can simply make the decision to buy the tools and get on with it.
>
> (Miller 2012)

For small specialist subcontractors and suppliers, the process of using digital technologies to facilitate data flows between design and production is not new. Niche market bespoke finishing joinery manufacturer Haldane UK's Managing Director, Forrester Adam, said:

> Since we bought our first 5-axis machine in 1992, we have been making steady improvements to our CNC techniques, however in 2009 we recognised that we needed a step change to take us to the next level. This required a radical change to virtually every process in the system and an investment to the tune of £300,000.

David Miller Architects (DMA) incremental methodology for BIM uptake	
Key action	**Reflection and follow-up**
In 2007 workstations converted to run BIM authoring software	Subsequent new employees have intensive training on BIM authoring software from Day 1
Trialled on small projects which could revert to 2D CAD if necessary	Reduced risk by incorporating backup
Live project £1m residential building using BIM authoring package to produce 2D drawings	Lonely BIM used as a precursor to Level 2 collaborative BIM
Process repeated several times on increasingly larger projects	Digital BIM information cross-referenced with latest NBS specification update
Some refurbishment projects had a tendancy to fall back into using 2D CAD	
In 2009 the practice committed to using BIM tools on all projects	In-house trainer and BIM Champion appointed.
First collaborative models produced for a live project	
Intensive six day training schedule developed. Broken down into 40 minute modules	Module structure allowed to fit training slot within the working day
BIM 'Boot Camp' introduced for new staff	Full immersion with no distractions during first week in office
Stability of working and training regimes for BIM implementation achieved	
BIM Champion role dropped and replaced with federated template where individual team members take responsibility for and explore different parts of BIM process	Office CPD sessions facilitate sharing of experiences and cross-pollination of ideas
Procedure for quantity take-offs developed in-house	
Furniture, Fixtures and Fittings procurement template developed. Linked to *Codebook* room data management software	Cross-platform data exchanges increased (author's note)
Operations and Maintenance (O+M) model developed for small student housing project	Potential to embed soft landings process into future project development (author's note)
First BIM Execution Plan written. In-house BIM manual and BIM Plan of Work written and linked to RIBA Plan of Work 2013 stages	Significant engagement with Level 2 BIM documentation (author's note)

Figure 13.1 Summary of David Miller Architects' BIM journey between 2007 and 2012

The drawing and programming aspect is one of the most critical and time consuming elements in the CNC process however this revolutionary technology powered by high powered processors allowed us to draw, programme and simulate the machining in a fraction of the time it previously took.

(Anon n.d.b)

The joined-up approach to using data incorporates high-definition surveying (HDS) laser technology for site surveys, which builds 'absolute precision' into the process to the extent that a full staircase and handrails can be manufactured from dry-jointed components for on-site assembly.

SME Cubicle Centre Marketing Director Craig Sewell noted that his company's BIM journey began after a meeting with leading BIM object developer and supplier BIMstore® in 2011. At that time, his specialist manufacturing/supplying firm became aware of BIM's potential not just as a marketing/specification tool, but also in offering possibilities to link design with manufacturing. With a company focus on efficiency, Cubicle Centre claims to be able to offer some of the shortest lead times in the industry, and BIM presented itself as an opportunity to make efficiency gains by improving workflows between design and manufacture.

Plus, internal company forecasts suggested that in the transition phase towards a fully integrated BIM process, by mitigating the risk of human error, the firm could reduce office-based production time by up to 33 per cent. These gains were effected through the use of data-rich BIM authoring software for design, and enhanced by the facility to harvest metadata from BIM models and link that digital data with the manufacturing process. There was also a significant awareness that grasping the nettle as early BIM adopters would allow Cubicle Centre to develop a competitive edge against much larger and more heavily resourced 'brand name' companies.

13.3 Maintaining internal and external quality standards

It is unlikely that a small organisation's first BIM project would be initiated from a blank canvas although, sadly, there is some anecdotal evidence from within the industry that some firms may be professing BIM competence and/or experience they are not able to deliver on in practice. That situation raises a number of questions. These range from issues linked with risk management to ethical and practical considerations. There has been a great deal of open discussion within the industry over the last few years on ownership of BIM data and professional indemnity insurance (PII) issues, particularly when data may be shared in collaborative digital environments. To date, there has been little compelling evidence offered by PII brokers that engagement with Level 2 BIM is likely to require onerous policy riders (or indeed any at all in some cases). As with many insurance matters, the key word may be 'disclosure' in dialogue between brokers and policyholders.

The Law Commission noted that much of the current law was governed by the Marine Insurance Act 1906, which codified principles developed in the eighteenth and nineteenth centuries. Although the 1906 Act only related to marine insurance, most of its principles were taken to reflect the law for all insurance. In their July 2014 report, the Law Commission and the Scottish Law Commission recommended reform of the law in four areas of insurance, including the duty of disclosure in business insurance. One of the recommendations was that insurers should take a more active role in the process of disclosure when assessing risk and not simply cite non-disclosure as a reason to reject claims.

Clearly, however, the premeditated and inaccurate completion of a prequalification questionnaire in pursuit of winning work would invariably invalidate PII cover. In that broad context, as BIM becomes more diffuse and pervasive within UK construction, verifiable evidence of organisational competence may become a more prevalent facet of professional practice. In that situation, the formal recording of BIM education, training and (in some cases) certification may assume a higher profile than at present.

The previous exemplars (Mann Williams and DMA) both suggested progressive and incremental methodologies for SME BIM implementation. Transitional phases bridged between conventional 2D CAD and model-based practices for information management. For both practices, BIM was required to dovetail with existing organisational structures, which had established protocols for quality standards. In that context, three issues which will invariably face SMEs and micros considering BIM implementation are industry and organisational standards for quality assurance, company accreditation and training.

These three facets of BIM practice may in some cases be mutually interdependent. An example of this would be where an SME or micro organisation was required to provide evidence of competence and experience as part of a prequalifying process (such as embedded in PAS 91:2013) for engagement as a partner in a Level 2 BIM supply chain. In terms of UK Government requirements for publicly procured projects, Level 2 BIM is currently ring-fenced by eight 'standards' (Figure 13.2). Rob Jackson of Bond Bryan Architects summarised these protocols succinctly in his BIM blog along with supplementary and cross-referencing quality management standards (Jackson 2014).

Ian Ritchie Architects Ltd (IRAL) published quality assurance standards which provide an exemplar of a consultancy with formal ISO 9001:2008 accreditation for quality assurance. The practice vision is:

> to provide a consistent, competent and professional service which is focused on satisfying the requirements and expectations of our clients and end users. The qualities which characterize this service are also reflected in our day-to-day relationships with professional colleagues and other members of the construction team.

That strategic purpose is articulated under three key headings:

- Management policy, which aims to embrace the quality management principles and requirements outlined in ISO 9001:2008 as a framework for improving performance to strive to create and maintain a humane work environment where staff are encouraged and given opportunities to continuously improve their skills.
- Environmental policy, which aims to embrace the principles of sustainable development in our designs.
- Health and safety policy, which aims to provide a safe working environment and to incorporate the best health and safety practices in the discharge of our duties as designers.

(Ritchie 2014)

The practice is able to demonstrate systematic, transparent and controlled workflows through implementation of the quality management (QM) system. It is also able to implement regular review and updating to ensure currency, effectiveness and relevance to practice requirements and enactment of a philosophy of continuous improvement. Central to that philosophy are the principles and practice of quality control as a key tool to ensure client and end user satisfaction. Quality assurance is a key principle underpinning information management and BIM practice.

Sometimes, embedding QM principles into organisational structures is a function of the client base and requirements. Ian Ritchie noted that his practice's work with the rail industry demanded significantly higher standards of quality management, consistent with meeting the relevant parts of BS50126:1999 'Railway Applications: The Specification and Demonstration of Reliability, Availability, Maintainability and Safety'. In order to harmonise with the protocol, IRAL implemented an ISO 9001-compliant quality management system with specific additional requirements to meet the Railway Industry Standards on transport infrastructure projects. The combined processes were subject to extensive audit by independent organisations

Eight pack suite of BIM protocols for UK Government publicly procured projects	
Standard	**Supplementary references**
PAS 1192-2:2013 Specification for information management for the capital/delivery phase of construction projects using Building Information Modelling	**BS EN ISO 9001:2008** Quality management systems **BS ISO 10007:2003** Quality management systems. Guidelines for configuration management **BS ISO/IEC 27001:2013** Information technology. Security techniques. Information security management systems.
PAS 1192-3:2014 (as updated) Specification for information management for the operational phase of assets using Building Information Modelling	**BS 1192:2007** Collaborative production of architectural, engineering and construction information *(purchase required).* See also Building information management – A standard framework and guide to BS 1192 **BS 1192-4:2014** Collaborative production of architectural, engineering and construction information – Client information requirements
BS 1192-4:2014 Collaborative production of information Part 4: Fulfilling employer's information exchange requirements using COBie Code of Practice	**COBie-UK-2012** (Construction Operations Building information exchange) http://www.bimtaskgroup.org/cobie-uk-2012/
PAS 1192-5:2015 Specification for security minded building information modelling, digital built environments and smart asset management (author's note: in draft at time of writing)	**BS ISO 15686-4:2014** Building Construction. Service Life Planning. Part 4: Service Life Planning using Building Information Modelling
CIC (Construction Industry Council) BIM Protocol First Edition 2013	http://www.bimtaskgroup.org/wp-content/uploads/2013/02/Outline-Scope-of-Services-for-the-Role-of-Information-Managment.pdf
Government Soft Landings (GSL)	**BSRIA the Soft Landings Framework** X-refers https://www.bsria.co.uk/services/design/soft-landings/
Classification Uniclass current version **Digital Plan of Work**	NBS digital **BIM Toolkit** X-refers http://www.thenbs.com/bimtoolkit/ or as current version

Figure 13.2 Schedule of published UK BIM–related protocols (after Jackson 2014)

prior to IRAL's formal accreditation as both architectural and lead consultants for London Underground Ltd.

Implementation of quality management systems has contributed to a continual improvement in the service that IRAL provides to its clients. A key feature of that process is that formalised quality assurance ensures that client, funder, stakeholder and user requirements are rigorously developed, recorded and monitored. Formal reviews are conducted by a senior member of staff during key project stages to verify project requirements. Technical and design submissions are signed off by a projects director. All project changes are rigorously managed by a change control process. The QM protocol also embodies processes which require continuous feedback both during the project development phases as well as post-handover.

Turning to accreditation, in the lead-in to the UK Government's BIM Level 2 gateway, many organisations (including professional bodies) have stepped up to the mark in offering BIM training. Whether (or not) training by an external provider is a necessary step along a BIM journey is for individual organisations to consider, reflect on and action as appropriate. There is certainly some evidence in the field that SMEs and micro firms are sufficiently proactive to be able to self-manage BIM training. Built environment professional bodies may require individuals to provide evidence of continuing professional development (CPD). BIM-related training can be formalised and embodied into a CPD record of activity.

Among providers, BRE Global has developed a BIM Level 2 scheme for BIM Certificated Professionals and offers a benchmark standard which is intended to underpin delivery of professional services. The BRE scheme offers certification that appropriately qualified BIM providers will meet or exceed threshold standards, through a combination of regular organisational audits, ongoing assessment and CPD recording.

In November 2014, BRE Global made it known publicly that international architecture and engineering practice BDP had become the first company to be assessed and certificated under BRE Global's BIM Level 2 Business Systems Certification Scheme. That announcement confirmed that BDP would have the policies and procedures required to deliver Level 2 BIM in line with the Government's strategy, meeting the requirements of PAS 1192-2:2013 and the PAS 91:2013 prequalification standard.

Feedback from the consultancy was that BDP was delighted to be the first organisation in the UK to have achieved BIM Level 2 Business Systems Certification for its London studio. Alistair Kell, Director of Information and Technology at BDP, commented that:

> this is a significant step along our BIM journey, both validating the investment and commitment we have made in redefining our processes to align with emerging BIM technologies whilst also providing a recognisable auditable standard to demonstrate our abilities externally. Our next step is to achieve Business Systems certification for the remainder of our studios and extend our employee capability through the BRE BIM AP accreditation scheme.

Among professional bodies, RICS currently offers a Certificate in Building Information Modelling – Project Management (equivalent to 200 hours CPD) which covers the entire BIM project life cycle and provides candidates with a detailed knowledge and required skill-sets to manage each step of a BIM project.

The RICS-certificated programme is underpinned by a simulated BIM project which runs across course delivery to embed practical knowledge at each stage of the BIM project process in developing a 12-point suite of learning outcomes:

1. identifying the case for BIM in any project, taking into account costs and benefits, and examining the major cost drivers, functionality and characteristics of a good BIM model;
2. creating a BIM execution plan, including the process, content, production and evaluation;
3. applying the processes and standards relevant to the whole life management of asset information;
4. identifying who and how to engage with the stakeholders at each stage of the project life cycle;

5. understanding the interaction of process, technology and people in a BIM environment;
6. applying the tools and techniques which support enhanced collaboration among the project stakeholders;
7. understanding the technology and Common Data Environment which supports BIM;
8. applying technology in one or more of the following: geospatial, design, cost, time and facilities management environments;
9. recognising the level of detail that BIM models can contain and how this relates to the stages of design, construction and maintenance;
10. knowing how BIM requirements can be implemented within the project's legal, procurement and tendering framework, and how to review examples in practice;
11. recognising the legal implications of BIM in terms of intellectual property, insurances and potential liabilities; and
12. evaluating contract requirements and commercial data of BIM models, and the inputs and outputs to be expected at each stage of the project life cycle.

While schemes like the RICS training package are helpful in identifying the broad scope of training required to get a BIM project up and running, some SMEs and micros may rely on less formalised methods of developing and attaining organisational competence with BIM. A number of the RICS learning outcomes may be already embedded in undergraduate syllabus structures (that point in itself raises the bigger issue of the strategic importance of pan-industry collaboration). SME and micro employers intent on self-managing the process towards BIM capability might find it productive to use graduate skills as a means of drip-feeding the development of BIM competencies within an office or organisational structure.

The National Federation of Builders (NFB) has been particularly active in encouraging its members to aspire to BIM capability and prepare accordingly by offering an holistic programme of activity supported by the Construction Industry Training Board (CITB). NFB also published an online diagnostic tool to test BIM capability and assess appropriate levels of support required by NFB and CITB members. Over the last three years, NFB's BIM Exemplar programme has provided intensive mentoring to 16 companies as they have upskilled for digital working. During that time, each company worked with an NFB BIM Adviser to assess the firm's BIM maturity in relation to the 2016 mandate and implement a structured programme to close knowledge gaps and encourage companies to be proactive with implementation.

Barnes Construction Ltd of Ipswich was one SME which tapped into the NFB Business & Skills' Building Information Modelling initiative. Following an introductory session hosted by NFB consultants, the Barnes board committed to setting up a detailed action plan with implementation over a 14-month period to prepare for delivery of construction projects to BIM Level 2 by 2016. Follow-up actions ranged from developing company policy and staff training to implementing new technology, processes and systems across their operations and supply chain, consolidated by BIM implementation on a pilot project.

13.4 Ethical, legal and professional liabilities

Purchasing BIM authoring and support packages plus training can represent a significant overhead for SMEs and micro organisations. One SME consultancy in Aberdeenshire reported taking BIM software on board, then reverting to 2D CAD after a few years because the firm came to the conclusion that BIM was not compatible with the practice's business model. In the case of the 30-employee Mann Williams engineering consultancy, director Spencer Fereday cited the capital expenditure to purchase one BIM authoring software licence as being around

£7k. Adding a desktop computer with sufficient random access memory (RAM) plus dedicated graphics card increased the cost of equipping one workstation for BIM to £8k. Sole practitioners from built environment disciplines may lack the resources to purchase software outright and may have to rely on cloud subscription services from software houses. That situation can introduce the risk of downtime in comparison with using software resident on a local hard drive. If the Internet or the provider's server is down, then users have no access to the BIM software.

There are also issues of currency (backwards compatibility with previous releases) and legitimacy of software to consider. SMEs and micros with stand-alone computers or operating networks with loosely controlled local networks (or in some cases where no effective management of computing resources exists) may be more vulnerable to breaches of cybersecurity than larger organisations with managed intranets. For example, an educational version of BIM authoring software might become embedded within a local network and be inadvertently used for commercial purposes. Software piracy (inadvertent or otherwise) is policed by the Business Software Alliance (BSA), which may carry out audits if it suspects an organisation of breaching software licensing conditions.

Research carried out by the Software Alliance in 2013 found that only 14 per cent of architecture firms are very confident the software in their firm is being used properly. It claimed 36 per cent of firms admitted illegal software could be in use at their company, and reported that just 9 per cent think intellectual property (IP) is valued within their company. The Alliance's director of compliance marketing, Julian Swan, said too many firms had a 'lacklustre approach to software management' and added: 'When a company's productivity, reputation and finances are at stake, software management should be taken seriously.' BIM processes are almost wholly dependent on storing and shifting digital data around virtual environments. Although employing j15 dedicated IT managers may be outwith the resources of some SMEs and micros, digital data still needs to be carefully managed and regularly backed up to avoid collateral damage from IT disasters, which could have serious financial consequences for a small organisation.

Within built environment practice, maintaining appropriate professional indemnity insurance (PII) could be regarded as a basic and universal tenet of business. Plus, the association of PII with business ethics assumes added significance when embedded within professional body codes of conduct. PII protects companies against expense incurred through defending financial claims made by clients or others. For example, beyond general consequences of mistakes being made in the course of business, loss of data, transmission of corrupted data and unintentional breach of data copyright are all BIM-related instances which might give rise to PII being invoked.

In 2011, Daniel Fierstein reported details of a claim which arose from a US university life sciences facility development where the project architect and MEP engineer had used BIM to trial build the MEP system into a ceiling plenum. During construction, the contractor was unaware that a specific installation sequence was necessary to get the equipment to fit, and ran out of ceiling space. Two lessons learned: first, BIM does not negate the need for good and regular communication across design and construction phases; second, construction contracts need to identify and reconcile the risks of using BIM so that roles and responsibilities are clearly defined from the outset.

Roxane McMeeken observed that the UK insurance industry's principal concern with shared digital data was 'confusion around who was responsible for what'. Following a series of consultations with the PII market, leading independent insurance broker Griffiths and Armour set out a best practice for PII in a guide first published in 2013 for CIC and the UK

Government BIM Task Group. The significant and common response from insurers was that use of discipline-specific federated models in a managed 3D digital environment (Level 2 BIM) was not a cause for concern when appropriate audit trails were in place. In fact, it was the potential robustness of these audit trails and systems for change management which raised confidence among the insurers. At this point, CIC guidance on PII is limited to Level 2 BIM and does not extend to Level 3, principally because of higher levels of complexity associated with a fully integrated Level 3 digital environment.

Koko Udom mapped a six-point structure in an NBS paper to scope out legal issues associated with BIM adoption:

1. the contractual framework within which BIM is incorporated;
2. model management;
3. intellectual property rights in parts or elements of the model;
4. reliance on data;
5. liabilities; and
6. ownership of BIM process and model.

(Udom 2012)

Udom argued that it was possible to establish that designing a building using BIM technology in collaborative environments raises certain specific legal problems which are not associated at all, or to such a large extent, with other types of design approach. In legal terms the most potentially problematic was the possibility for disaggregation of design responsibility. In a broader context, management of the BIM process, and the interplay between the theoretical clarity of the process, versus its relative messiness in practice, may pose legal and organisational challenges. The most significant of these according to Koko Udom is establishing the best way of ordering the contractual and organisation structures to give the process maximum support. That process needs to evolve by consensus between the commissioning client and the design-construct team and procedures embedded in the key components of BIM documentation, the Employers Information Requirements (EIR) and BIM Execution Plan. If necessary, these arrangements can be bound into contractual arrangements using the 2013 CIC BIM Protocol.

13.5 Protocols and structures for file management: BS1192:2007

In computing, a directory is a component of a cataloguing hierarchy which contains references to digital files. In the interests of fast and efficient manipulation of data, computer operating systems demand structured and organised protocols for the generation and management of folder and file structures. The use of desktop folder icons as metaphors for computer operating system directories has been ubiquitous for many years, and the conventions cut across industry and professional discipline boundaries. In the 1970s, within the broad spectrum of UK architectural education, basic programming skills were taught, and some contemporary BIM theoreticians and digital process-masters may have benefited from that experience of structured and methodological working in dealing with digital data.

During the 1980s, practitioners could script virtual building animations from captured 3D CAD views using Autolisp, the dialect of Lisp programming 'built' for use with AutoCAD and its derivatives. Like G-code, the language used to tell computerised machine tools (latterly 3D printers) how to make things, Lisp originated in the 1950s for artificial intelligence applications and is still in use. Baby boomer and Generation X practitioners may have memories of creating games and animating them with sprites using the STOS language and

Atari ST home computers. Hard drive and graphic capabilities may have moved on, but core principles are still applicable. As sophisticated BIM authoring packages started to enter the marketplace, some software offered users the facility to customise and script their own objects. From around 1998, David Nicholson-Cole's work with Graphisoft ArchiCAD and the GDL language encouraged bridge building between the more arcane aspects of computer science and practicalities of designing/prototyping buildings and their constituent parts in 3D virtual environments.

Microsoft's plug and play paradigm evolved into an information technology game changer. The need for computer users to have underlying knowledge and understanding of file and folder protocols gradually became diluted and redundant. Accompanying that obsolescence came a loss of control and increased reliance on specialists to troubleshoot problems and develop new knowledge and supporting technologies. In a contemporary setting, augmented reality (Figure 13.3) is on the crest of that socio cultural shift towards dependency on digital media to service the activities of daily living and the workplace.

Technology never stands still; one of the biggest challenges in contemporary workplace environments is having the facility to capture data and fit it into workflows at appropriate points. In that context, BIM at a project level seems considerably down the information food chain and subordinate to much larger aggregations such as 'big data' and Geographic Information Systems (GISs). The layering, accessibility and control of data has become more of an issue in recent years. Data has also developed a nuisance characteristic which may require filtering, for example in the separation of junk email or the invasion of screen environments with unwanted and superfluous advertising.

Within the wide spectrum of activity defined by digitisation generally, and as applied to design, construction and building maintenance/management processes, the need to be able to locate specific files quickly and efficiently is more important than ever. Anyone who has covered an absent co-worker's unfamiliar file structure via a desktop PC will know that. File organisation is key to fast and efficient retrieval of data. Not as quickly as would happen on a computer hard drive, but the principle applies universally whatever the size of the organisation. Also, file identification and referencing is key to the send/receive aspects and location/interpretation to the most up-to-date data available. Digital file-related issues, which may not be hugely important for sole practitioners, will inevitably assume greater significance when bad information is posted to and shared within design and construction teams.

One practical consequence of BIM is that data is generated, aggregated, modified and moved around in digital environments. Self-evident possibly, but worth reflecting on some of the implications for practice, particularly when project-based digital environments are shared by a number of users:

- aggregation suggests that graphical and alphanumeric information (referred to as meta-data when associated with BIM objects), previously spread between drawings, specifications and bills of quantities, will be embedded within BIM models;
- that may raise issues of visibility, granularity (the extent to which a BIM environment is comprised of distinguishable pieces) and accessibility, particularly for non-expert users;
- data in transit flags up caveats which apply universally to digital media. From a practitioner's perspective, these can be fragile and transient information storage containers.

Pre-emptive and careful husbandry of tools for generating, storing and manipulating digital information may be particularly significant for SMEs and micros working in stand-alone PC environments or small self-managed networks. At project level, the use of BIM taxonomies

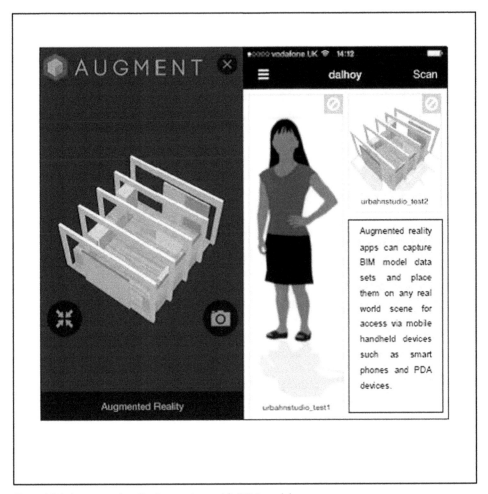

Figure 13.3 Augmented reality interactions with BIM models

for structuring and ordering data are important, but can be unwieldy for small flat organisations lacking the resources or inclination to apply the rules to their portfolios of projects. In that context, the COBie protocol is often raised in practitioner discussions as a case in point.

The British Library has roots in the British Museum's Department of Printed Books, founded in 1753, and currently holds around 170 million items. That's around one-third of tweets currently being processed by Twitter's servers in a 24-hour period. These are the two extremes of data storage and management. Imagine entering the British Library on a quest for a particular book chapter or journal article. Without access to a hierarchy of classifications, titles and contents, a search would not be feasible. Similarly, retrieval of digital information quickly and efficiently in a business environment demands a structured approach. Bespoke systems for information capture may work well within a small company, but when digital data needs to be generated, manipulated and viewed between individuals and across organisations, some level of standardisation becomes a necessity to avoid at best unproductive time, confusion and ultimately chaos.

The Building Project Information Committee (BPIC), set up in 1987 under the joint sponsorship of RIBA, RICS and others, launched the CPI suite of documents for project co-ordination. BPIC morphed into CPIC and 20 years on the British Standards Institution published a new code of practice for collaborative construction information. BS1192:2007 incorporated work methods established and refined in the Avanti programme initiated by the DTI in 2002. Avanti's primary objective was to deliver improved project and business performance through the use of digital tools which supported collaborative working. In 2006, the Avanti DTI documentation was transferred to the Constructing Excellence organisation.

BS1192:2007 applies equally to the management of 2D drawing information and BIM projects. As the cornerstone and first base on a journey towards compliance with the Government's 2016 BIM target, the standard defines three specific areas which must be addressed to enhance the production information process: roles and responsibilities, Common Data Environment (CDE) and Standard Method and Procedure (SMP). More on BS1192:2007 follows later.

13.6 Practical aspects of data sharing and exchange

Data protection and privacy laws impact on most businesses. The issues they raise can be both complex and pervasive. Penalties for non-compliance are becoming increasingly punitive. Plus, the burgeoning exploitation of big data and open data provides commercial opportunities which need to be handled ethically and within the law.

It is becoming increasingly difficult to manage external data interventions within the activities of daily living including in the workplace. Even with industry standard virus protection and firewalls embedded on desktop hardware and mobile devices, it is still sometimes very difficult to distinguish 'good' and useful information from the unwanted and in some cases potentially dangerous data which roams around the digisphere looking for internet protocol (IP) addresses to invade. BIM projects held in the cloud or on multi-access servers are vulnerable to hackers and data corruption in a way that CAD drawings resident on local hard drives never were.

In February 2015, BSI released the draft PAS 1192-5 for public review and comment. PAS 1192-5 is stated to provide 'a framework to assist asset owners and shareholders in understanding the key vulnerability issues and the nature of the controls required to enable the trustworthiness and security of digitally built assets'. The document was not designed to hinder collaboration in the name of protecting sensitive data but rather to 'ensure that information is being shared in a security minded fashion'. So the document addresses the need to be security minded with BIM, in particular consideration being given to cybersecurity.

Cybersecurity is a means of protecting digital domains such as interconnected networks and computer-based systems. BIM relies on collaboration and many iterations of data, which may be stored in the cloud or on a BIM server. By definition, the Common Data Environment (CDE) outlined in BS1192:2007 facilitates multi-user access. The procedures put in place to protect the information held in CDEs are key components of establishing and maintaining trustworthy BIM environments. For SMEs and micro organisations, an essential precursor to engagement with BIM projects is a robust internal review of data protection and potential digital data security issues, particularly those likely to impact on collaborative working.

The Institution of Engineering and Technology (IET) published a code of practice for built environment cybersecurity, including the identification/management of risks which are inherent in the adoption of BIM models, collaborative processes and systems. The Code examined different sources of threats across the building life cycle from initial concept through to decommissioning, and addressed threats to intellectual property, commercial data and the design/operation of building systems. The IET's checklist of pre-emptive actions applicable to BIM teams includes:

- understanding cybersecurity;
- developing a cybersecurity strategy;
- using trustworthy software;
- configuration control;
- managing process and procedures;
- managing people aspects;
- managing technical aspects;
- developing a cybersecurity policy for a building; and
- applying cybersecurity across the life cycle of a building.

The SANS Institute produced a 20-point list of critical security controls to protect digital working in commercial environments. In 2013, the stewardship and sustainment of these controls was transferred to the Council on CyberSecurity, an independent, global non-profit entity committed to a secure and open Internet.

openBIM® is a universal approach to the collaborative digital design/operation of buildings based on open standards and workflows. The buildingSMART® alliance is the international organisation with a mission to develop and propagate the use of open BIM standards in collaboration with software houses and the construction industry globally. In particular, buildingSMART® has been involved with evolution of the IFC file type protocol. IFC2x3 (IFC4 is in prototype form at the time of writing) is the 'universal' standard developed to facilitate file transfer across different software platforms and between architectural, structural and building services engineering BIM models (Figure 13.4).

As the IET Code identified, one aspect of taking an holistic approach to cybersecurity is the issue of software verification, and buildingSMART® publishes a list of BIM-related commercial programmes which have been tested and certified to support the IFC file transfer process. The Green Building XML (gbXML) open schema is a file format commonly used to export data from BIM authoring platforms and into specialist energy analysis software such as the Integrated Environmental Solutions (IES) suite. The gbXML organisation publishes online file verification software which users can test drive to check the robustness of gbXML model files before sharing files across the design-construct team.

Moving from pan-industry to project-specific considerations, the CDE is at the heart of any collaborative BIM project. In developing the commentary introduced in Section 13.5, the CDE will typically exist as a master folder on a BIM server or as hosted within a cloud accessible digital environment. In terms of the issues discussed, bearing in mind that a CDE (or its variants) may serve project requirements over design/construct/occupancy phases, the need for robust measures to be put in place to protect the integrity of the CDE cannot be over-emphasised.

Mervyn Richards' guide to BS1192:2007 (Richards 2010) is the definitive document and essential reading to aid navigation through the structuring, protocols and conventions associated with developing a CDE which harmonises with the PAS 1192-2 BIM standard. BS1192:2007 is not BIM exclusive; many of the procedures described are equally applicable to information management for projects using 2D CAD. The principles outlined in Richards' handbook are universal and apply whatever the scale of the project. Underpinning practical

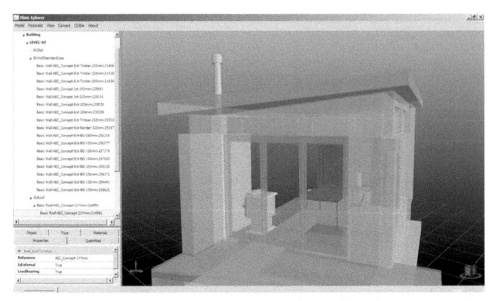

Figure 13.4 BIM model authored in Autodesk Revit and saved in IFC format for viewing using xBIM
Xplorer freeware developed by BIM Academy/Northumbria University

aspects of BIM engagement is the central premise that the CDE provides an environment for 'iterative development of the design documentation to achieve full integration and spatial co-ordination of the data/information from all participants and offices, and from all originators within project supply chains' (Richards 2010).

The BS1192:2007 handbook does not cover the specifics of using digital media to facilitate online collaboration between design-construct team members. These interactions can be structured and executed using proprietary design or project management software packages. The CDE would typically sit within the preferred software environment. These managed collaborative tools effectively act as password-protected business-to-business extranets. The wide variety of available packages ranges from ubiquitous freeware like Asana (suitable for entry-level engagement), through mid range multipurpose applications like Basecamp, to specialist built environment digital tools such as offered by 4Projects, Asite and similar Software as a Service (SaaS) providers.

The 'high-end' packages may be able to record user activity in some detail, for example tracking changes to files logged on the host server. These tools may be task driven and are structured to allow online dialogue between participants in real time; a process more robust to audit than communication by email. Some of these software offerings (particularly at the 'lower end' of the affordability range) are essentially social media derivatives tailored for commercial environments.

At set-up (BIM kick-off meeting), and prior to authoring the BIM Execution Plan (BEP), it is essential that protocols for the structuring of BIM models and CAD drawings (as applicable) are agreed between partner consultancies. If the conversations highlight different work practices/conventions, some harmonisation may be required by the project information manager. Ownership of data within the CDE, including discipline-specific models, remains with the originating organisation. Data which the client requires as deliverables from the CDE at defined milestones during project development will need to be identified and incorporated into the Employer's Information Requirements (EIRs).

These actions will put down significant procedural markers at the project briefing phase. In the case of a procurement route which involves early contractor/subcontractor/supplier engagement, the design-construct team may be collaborating from the first stages of project evolution. In the interests of consistency of structure, content and presentation of data, the whole team needs to buy into these procedural conventions, which will be embedded in BIM documentation.

In his guide to BS1192:2007, Mervyn Richards listed eight principles for the development and presentation of robust design and construction information. These are condensed into four key points for discussion, consensus and recording by the project team:

1. file naming/annotation conventions, including reference codes for status and revision of files;
2. project GIS origin based on real world co-ordinates. For example as referenced to the Ordnance Survey (OS) National Grid. A common origin point will be required to synchronise discipline-specific models referenced from within viewing and clash detection software. Project datum for vertical levels to be established;
3. drawing sheet templates, including title blocks, text fonts, paper sizes and production scales;
4. layering conventions (where applicable) based on BS1192 and classification protocols. The default is Uniclass (current update). Some harmonisation may be required between discipline-specific BIM authoring packages, which use different classification conventions.

The Common Data Environment (CDE) serves two key purposes: to act as a repository for collaborative model data and to host project deliverables, such as 2D drawing files saved as model views. The scope of the CDE needs to be defined in the BEP. In particular, protocols for file sharing, file management between disciplines, and recording input to and output from digital file inventories need to be articulated. In applying the BS1192:2007 template, the CDE is divided into four areas (folders) for operational purposes. Each area will be represented by a separate folder; each folder represents work at a different stage of development. The AEC (UK) BIM Protocol V2 provides exemplar subfolder structures.

At project level, the four activity folders are 'Work in Progress', 'Shared', 'Published' and 'Archive'. Individual folders are populated by files appropriate to developing each activity. One of the phenomena which characterise BIM projects is the diversity of file types which may populate the CDE for a typical project. Information containers associated with these different file types need to be structured consistently and stored across the discipline-specific folders.

Typical file types will include native model files, IFC format for export and 3D model viewing between disciplines, and DWG, DWF, DXF formats for 2D CAD drawings, Adobe PDF files (commonly used for deliverables like contract drawings), image files JPG, GIF, TIF, PNG, Excel spreadsheet files (XLS) for COBie data, and environmental analysis files in native or neutral (for example, gbXML) formats. Even a small project might involve cross-disciplinary dialogue which draws from an inventory of up to 20 different source file types. As a subset of file types, there is also the issue of versions, for example, the IFC file format has evolved through a number of versions. BIM authoring software providers may use different variations of 'standard' file types. It is a complex digital mix which needs to be managed carefully and with authority.

The designated information manager has the key role of 'gatekeeper' in monitoring the signing-off process, as data moves from one folder to the next, and in controlling information flow between folders.

The 'Work in Progress' (WIP) folder represents individual discipline activity. The WIP area of the CDE will be subdivided into discipline-specific folders where members of the project team log their own work. Typically, these folders will include model files in native, IFC file and possibly other formats as described previously. The most commonly used BIM authoring software packages all use different native file formats. The IFC protocol was intended to offer an agnostic file type which would permit 'dialogue' between different BIM authoring software packages in the spirit of openBIM®.

Each discipline may subdivide its master folder for the purposes of structuring information, for example to distinguish between BIM models and 2D CAD drawings required for statutory approval submissions or contract purposes.

Establishing conventions for file numbering is important and these need to be applied consistently throughout the CDE. BS1192:2007 offers guidance on appropriate numbering schema for model (including object models) and CAD files. The 'Work in Progress' folder must also incorporate the facility for updating files.

The 'Shared' folder represents a repository for cross-disciplinary activity. To facilitate co-ordinated, efficient working, each contributor to the CDE makes 'own data' available for shared access across the design and construction team in the 'Shared' folder. In the situation where proprietary design/project management software is being used, these files would normally be accessible to all team members with login rights. Before uploading model files, individual disciplines need to check, validate and sign off data sets in compliance with BS1192:2007 recommendations and 'own organisation' quality assurance protocols where applicable.

Sharing of discipline-specific models and files (IFC, gbXML etc.) needs to be carried out on a regular basis in order that all disciplines are working to latest validated data as defined in the project BEP. The 'Shared' folder also functions as a repository for external provider data uploaded for sharing across the project team (for example, infrastructure and services information from statutory authorities).

On specific projects, the designated information manager is responsible for change management and needs to ensure that the 'Shared' folder is updated regularly (for example, weekly). All changes to shared data should be systematically logged and communicated to the project team. Procedures for ensuring that process is effective, transparent and auditable need to be embedded in the BEP.

In essence, there is nothing new about many of these process controls. They exist to ensure that shared data has been validated by individual disciplines within their organisational quality assurance frameworks, is fit for purpose and current at the point of access. That in itself is a risk management process and was a well-trodden path before BIM. It is principally the operational environment which is different.

A fresh challenge which BIM brings to the frame of reference is that (compared with 2D drawings) BIM models will invariably be rich in geometric and alphanumeric metadata. But that data is not always visible, and readily accessible to team members, particularly non-experts like clients and building users. The information manager holds a key role in managing the process of ensuring that data is readily available when required and in a form which can be read and understood by the receiving party.

The 'Published' folder represents validated data which has been signed off by the client and is suitable for sharing outwith the project team. Published information refers to documents and other deliverables generated from information held in the 'Shared' folder. Typically, this will include exported data, drawings prepared for planning and building warrant submissions, tender drawings, contract drawings, reports and specifications.

The 'Archive' folder represents information which will be transferred from the Project Information Model (PIM) to the Asset Information Model (AIM) which serves post-handover information requirements. Information which has been updated or superseded can be logged to one subfolder of the 'Archive' master folder. A second subfolder can contain information required post-handover, such as record of protocols used for transferring signed-off project information from the PIM to the AIM, change audits and asset registers, discipline-specific and pan-discipline federated models, documents to support operational functions and statutory/legal requirements (operational fire safety arrangements, Health and Safety file) operation and maintenance information, maintenance manuals.

Complying with the procedures embodied in generating and maintaining a structured CDE demands a disciplined, rigorous and consistent approach to file management. All design and construction team members need to buy into and apply that ethos. Since the early 1990s, industry initiatives and reports have demonstrated that UK construction has a poor track record in producing accurate, unambiguous and complete production information. That situation impacts on the time, cost and quality metrics used to measure the quality of output from design and construction teams.

BS1192:2007 offers a robust template for information production and sharing which harmonises with the PAS 1192-2 and PAS 1192-3 BIM protocols. The responsibility for applying these standards lies firmly in the hands of the industry at every organisational level, from micro-SMEs and SMEs through to the largest construction firms and project-based supply chains. Taking a loose and ad hoc approach to data sharing will not work with BIM. Even small projects can involve the manipulation of large, complex and diverse data sets. The rules of engagement for the management of digital data have to be set out clearly from project inception and consistently followed through the design, construct and post-occupancy phases.

Bibliography

AIA (2007) *Integrated Project Delivery: A Guide*, The American Institute of Architects, Version 1, [online]. www.aia.org/groups/aia/documents/pdf/aiab083423.pdf [accessed 16 December 2014].

Anon (2005) *The New SME Definition, User Guide and Model Declaration*, European Commission, [online]. http://bookshop.europa.eu/en/the-new-sme-definition-pbNB6004773/ [accessed 26 March 2015].

Anon (2008) *What is Avanti?*, Constructing Excellence, [online]. www.constructingexcellence.org.uk/ceavanti/about.jsp [accessed 19 December 2014].

Anon (2009) 'Organisational Agility: How Business Can Survive and Thrive in Turbulent Times', Economist Intelligence Unit, *The Economist*, [online]. www.emc.com/collateral/leadership/organisational-agility-230309.pdf [accessed 16 December 2014].

Anon (2011) *Future-Proofing BIM*, Dell and BD+C, [online]. https://www.wbdg.org/pdfs/1103_dell_bdc_whitepaper.pdf [accessed 16 December 2014].

Anon (2012a) *AEC(UK) BIM Protocol, Version 2*, AEC UK, [online]. https://aecuk.files.wordpress.com/2012/09/aecukbimprotocol-v2-0.pdf [accessed 29 December 2014].

Anon (2012b) *AutoLISP Developers Guide*, Autodesk, [online]. http://docs.autodesk.com/ACDMAC/2013/ENU/PDFs/acdmac_2013_autolisp_developers_guide.pdf [accessed 16 December 2014].

Anon (2013) 'Partners', *BIM Partners*, Building Information Modelling (BIM) Task Group, HM Government Department for Business Innovation and Skills, [online]. www.bimtaskgroup.org/bim4smes/ [accessed 24 March 2015].

Anon (2014a) 'BDP Achieve a BIM Level 2 Certification First', *News from the BRE Group*, BRE Global, 11 November, [online]. www.bre.co.uk/news/BDP-achieve-a-BIM-Level-2-certification-first-1025.html [accessed 24 March 2015].

Anon (2014b) *Insurance Contract Law: Business Disclosure; Warranties; Insurers' Remedies for Fraudulent Claims; and Late Payment*, Law Commission, 17 July, [online]. http://lawcommission.justice.gov.uk/publications/insurance-contract-law.htm [accessed 26 March 2015].

Anon (2014c) *Quality Plan, BIM Level 2 Business Systems Certification Scheme*, BRE Global, [online]. www.bre.co.uk/filelibrary/accreditation/scheme_documents/BIM/LP527.pdf   [accessed   29 December 2014].

Anon (2014d) 'Smart Schools – David Miller Architects', *2014 London Design Awards*, The Design 100, [online]. http://londondesignawards.co.uk/lon14/entry_details.asp?ID=13720&Category_ID=5922 [accessed 24 March 2015].

Anon (2014e) 'Welcome', gbXML.org, [online]. www.gbxml.org/ [accessed 24 March 2015].

Anon (2015a) 'Critical Security Controls', SANS Institute, [online]. https://www.sans.org/critical-security-controls/ [accessed 24 March 2015].

Anon (2015b) 'Data Protection and Privacy', Pennington Manches, [online]. www.penningtons.co.uk/expertise/solicitors-for-business-data-protection-and-privacy/ [accessed 23 March 2015].

Anon (2015c) 'NFB Recognises Barnes as BIM Exemplar Company', *News*, Barnes Construction Ltd, 2 March, [online]. www.barnesconstruction.co.uk/News/Detail.aspx?id=95 [accessed 26 March 2015].

Anon (2015d) 'Sir Geoff Hurst MBE Congratulates NFB BIM Exemplar Companies', *Latest News*, National Federation of Builders, 19 February, [online]. www.builders.org.uk/nfb11/News.eb?nid=330318 [accessed 26 March 2015].

Anon (n.d.a) 'About BSA', The Software Alliance, [online]. www.bsa.org/about-bsa [accessed 18 March 2015].

Anon (n.d.b) 'Haldane Celebrate 20 Years in CNC Machining with Revolutionary New Approach', Haldane, [online]. www.haldaneuk.com/news/201202/haldane-celebrate-20-years-cnc-machining-revolutionary-new-approach[accessed 24 March 2015].

Anon (n.d.c) 'Software Certification', buildingSMART®, [online]. www.buildingsmart.org/compliance/software-certification/ [accessed 24 March 2015].

Brister, A. (2014) 'Cyber Threat', *Architectural Technology*, Issue 112, CIAT.

BSA (2014) 'Campaign 2014', BSA: The Software Alliance, [online]. www.bsa.org/anti-piracy/eufairplaycampaign [accessed 23 December 2014].

Chambers, T. and Wood, J.B. (1998) 'Information Technology in the Building Design Engineering Studio', in: *Computer Craftsmanship in Architectural Education*, eCAADe Conference Proceedings, Paris, France, [online]. http://cumincad.architexturez.net/system/files/pdf/40db.content.pdf [accessed 18 December 2014].

CIC (2013) *Best Practice Guide for Professional Indemnity Insurance when Using Building Information Models*, Construction Industry Council, [online]. www.bimtaskgroup.org/wp-content/uploads/2013/02/Best-Practice-Guide-for-Professional-Indemnity-Insurance-when-using-BIM.pdf   [accessed   23 December 2014].

Fenby-Taylor, H. (2015) Face-to-face comment made at BIM4SME meeting, Institute of Civil Engineering, London, November 2015. Subsequently blogged at www.bim4sme.org/news/blogs/being-inspired-by-bim4sme/ [accessed 19 August 2015].

Fereday, S. and Potter, M. (2013) 'From Lonely BIM to Design for Manufacture and Assembly (DfMA): A Learning Curve for One SME', *The Structural Engineer*, 91(11, November), [online]. www.istructe.org/journal/volumes/volume-91/issues/issue-11 [accessed 22 December 2014].

Fierstein, D. (2011) 'World's First BIM Claim', *Construction Law Signal*, 22 June, [online]. www.constructionlawsignal.com/by-subject/design-and-technology/worlds-first-bim-claim   [accessed   23 December 2014].

Fried, J. (2011) 'Why I Run a Flat Organisation', *INC* magazine, April, [online]. www.inc.com/magazine/20110401/jason-fried-why-i-run-a-flat-company.html [accessed 16 December 2014].

Hammad, D.B., Rishi, A.G. and Yahaya, M.B. (2012) 'Mitigating Construction Project Risk Using Building Information Modelling' (BIM), *Proceedings of 4th West Africa Built Environment Research (WABER) Conference*, Abuja, 24–26 July, [online]. www.academia.edu/2011136/MITIGATING_CONSTRUCTION_PROJECT_RISK_USING_BUILDING_INFORMATION_MODELLING_BIM_ [accessed 24 March 2015].

IET Standards (2014) *Code of Practice for Cyber Security in the Built Environment*, The Institute of Engineering and Technology, [online]. www.theiet.org/resources/standards/cyber-cop.cfm [accessed 24 March 2015].

Jackson, R. (2014) 'Standards', *BIM Blog*, Bond Bryan Architects, [online]. http://bimblog.bondbryan. com/standards/ [accessed 24 March 2015].

Lepage, M.R. (2013) 'A New Business Model for Small Firm Architects', *Entrepreneur Architect*, [online]. www.entrearchitect.com/2013/10/13/a-new-business-model-for-small-firm-architects [accessed 18 December 2014].

Lester, S. (2009) 'Routes to Qualified Status: Practices and Trends among UK Professional Bodies', *Studies in Higher Education*, 34(2), Routledge, [online]. www.sld.demon.co.uk/pqroutes.pdf [accessed 16 December 2014].

Levy, F. (2012) *BIM in Small-scale Sustainable Design*. USA: Wiley.

McAdam, B. (2010) 'The UK Legal Context for Building Information Modelling', *Proceedings, W113 – Special Track, 18th World Building Congress*, Salford, UK, [online]. www.lawlectures.co.uk/w113/documents/wbc2010-proceedings.pdf [accessed 29 December 2014].

McMeeken, R. (2012) 'Insurance Special Report: BIM', *Building*, [online]. www.building.co.uk/insurance-special-report-bim/5044731.article [accessed 23 December 2014].

Miller, D. (2012) 'BIM from the Point of View of a Small Practice', Building Information Modelling, NBS, [online]. www.thenbs.com/topics/bim/articles/bimsmallpractice.asp [accessed 29 December 2014].

Nicholson-Cole, D. (2000) *The GDL Cookbook 3*, Marmalade Graphics, [online]. www.nottingham. ac.uk/~lazwww/cookbook/CB_download/Cookbook3_1.pdf [accessed 18 December 2014].

Novak, J. (2014) 'The Six Living Generations in America', *Marketing Teacher*, [online]. www.marketingteacher.com/the-six-living-generations-in-america [accessed 17 December 2014].

Ogunjemilua, K., Davies, J.N., Grout, V. and Picking, R. (2009) 'An Investigation into Signal Strength of 802.11n WLAN', *Proceedings of the Fifth Collaborative Research Symposium on Security, E-Learning, Internet and Networking*, Darmstadt, Germany, [online]. www.glyndwr.ac.uk/computing/research/pubs/sein_odgp.pdf [accessed 18 December 2014].

Ramchurn, R. (2014) 'AJ Spec Live: Structure and Services', *AJ*, 23 May, [online]. www.architectsjournal.co.uk/news/aj-spec-live-structure-and-services/8663064.article [accessed 14 January 2015].

Richards, M. (2010) *Building Information Management. A Standard Framework and Guide to BS 1192*, BSI Group.

RICS (2014) 'Certificate in Building Information Modelling (BIM) Project Management', *Training and Events*, Royal Institution of Chartered Surveyors, [online]. www.rics.org/uk/training-events/e-learning/distance-learning/certificate-in-bim-implementation-and-management/online/ [accessed 29 December 2014].

Ritchie, I. (2014) 'Quality Assurance', Ian Ritchie Architects Ltd, [online]. www.ianritchiearchitects.co.uk/profile/quality-assurance/ [accessed 29 December 2014].

Rock, S. (2015) 'PAS 1192-5 Draft Published: Can We Collaborate on Cyber Security?', *BIM+*, CIOB, 25 February, [online]. www.bimplus.co.uk/management/can-we-collaborate-cyber-security/ [accessed 25 March 2015].

Rogers, D. (2013) 'Architects Warned over Software Infringements', *Building*, 11 January, [online]. www.building.co.uk/architects-warned-over-software-infringements/5048355.article [accessed 24 March 2015].

Udom, K. (2012) 'BIM: Mapping out the Legal Issues', Building Information Modelling, NBS, [online]. www.thenbs.com/topics/bim/articles/bimMappingOutTheLegalIssues.asp [accessed 23 December 2014].

WYLLN (2010) *A Guide to Professional Institutions Within Construction and the Built Environment Sector*, West Yorkshire Lifelong Learning Network, [online]. www.lcb.ac.uk/pdfs/WYLLN.pdf [accessed 19 October 2014].

14 Interpreting and applying PAS 1192-2:2013

14.1 PAS 1192-2:2013 as an industry model

Since publication of the Government Construction Strategy in 2011, the Building Information Modelling (BIM) Task Group have supported and helped deliver on central government's objectives, in particular the requirement to strengthen the UK public sector's capability with BIM implementation. The BIM Task Group offered a straightforward hypothesis to the UK construction sector. Significant gains in cost, value and carbon performance could be achieved through increased use of open sharable digital data for making/using/maintaining building and infrastructure projects. To what extent that premise can be validated may as an industry model become clearer over the next 5–10 years. In the meantime, as a starting point, the UK BIM Task Group's key operational intent was to achieve collaborative BIM Level 2 maturity on all central government department procured projects by 2016.

The primary function of the PAS 1192-2:2013 document is both to support the Government's aspirations with digitisation of construction and provide more general industry guidance in setting out a framework for collaborative working and information management on BIM-enabled projects. As far as the BIM Task Group is concerned, PAS 1192-2:2013 and its sister document BS1192:2007 are mutually interdependent as the only tried and tested standards which support the Government's BIM strategy to achieve Level 2 compliance.

Two years on from publication of the UK Government's 2011 Construction Strategy, the genesis and evolution of PAS 1192-2:2013 as a compliance benchmark marked an industry watershed for the management of digital data. The essence of PAS 1192-2:2013 was to present construction UK with a template for doing collaborative BIM. Most significantly, the guidance set out a matrix of key principles and practical actions for applying Level 2 BIM to projects. The PAS authors argued their document offered equal value to small practices and large multinationals. It is also important to note that PAS 1192-2:2013 sits as one standard among a comprehensive suite of BIM-related documentation most recently supplemented by the NBS BIM Toolkit.

Although development of the PAS suite was informed by a consultative process, there is little published evidence that the draft standards were road-tested in the field among SMEs and micro organisations. In some ways, mapping out UK construction's response to the PAS suite is like a frontier-land: the territory is largely uncharted and there is little tradition or knowledge base to draw from for small projects. Two schools of thought have emerged, mainly through blog commentaries and online discussions. The first is that the structure of PAS 1192-2:2013 incorporates some flexibility of interpretation when applied to building and infrastructure developments. The second suggests that if the PAS is not used 'as is', different interpretations across the industry will result in dilution of meaning and lack of consistency in project-specific applications.

Without some evidence-based body of knowledge to draw from, it is difficult to predict how the industry response to the PAS suite will pan out over the next ten years. At the time of writing, the PAS 1192-5 protocol was still at draft stage. However, the PAS 1192-2:2013 template certainly poses some questions in relation to its application. For example, will the PAS prove to be an industry game changer, or does it represent a reference point which SMEs, micros and their commissioning clients may be aware of, but feel no need to engage with? Will PAS 1192-2:2013 assist small organisations in migrating to collaborative BIM or does it present more challenges than incentives? Some SMEs and micros have expressed a fear factor with BIM adoption. Is that a fear of BIM per se, or of change, or of the machinations which collaborative BIM introduces into the design/develop/construct process, or possibly aspects of all three?

From a reader's perspective, PAS 1192-2:2013 is not a hugely challenging document to navigate through, but it does introduce roles and administrative layers for information management which may not necessarily fit with small project requirements or methods of working. While it has been argued that consequences of poor information management have plagued UK construction for years, the Government's focus in driving the digital technologies agenda forward has clearly been on public sector asset procurement. Despite the stated universality of the PAS 1192-2:2013 standard, it would be naïve to pretend otherwise. In fact, the Government's intent on that score is manifest throughout the PAS 1192-2,3,4 suite of documents now in the public domain.

Consequently, in posing the question 'will those within the industry with less dependency on publicly procured projects be playing catch-up with early adopters?', on the evidence available so far, the likely answer is, 'yes, it is certainly looking that way'. In an already diverse and fragmented sector, the prospect of further subdivision into 'BIM-smart' and 'BIM-dumb' sectors of activity is not an appealing prospect. Plus, it could be argued the Government's 2025 vision for a digitised construction sector requires wholesale industry buy-in to be deliverable.

As with the overarching relevance of PAS 1192-2:2013 as an industry protocol, the merits of BIM as a performance enhancing agent have yet to be supported by significant analytical data from across the construction supply chain. That situation has proved to be a moot point for some observers. For example, although not defined from the outset as a BIM project, it was reported in 2013 that key elements of Dundee's V&A Museum of Design (structure + services) had been developed in co-ordinated 3D data-rich environments which enshrined the key principle of digital prototyping. Yet in January 20015, news reached the public domain that capital costs for the new building had spiralled from £45 to £80 million. On the other hand, the gargantuan £842 million South Glasgow University Hospital (SGUH), which commenced in 2010 (predating publication of PAS 1192-2:2013 by three years), was handed over ahead of schedule and on budget by global contractor Brookfield Multiplex Construction Europe (BMCE). To facilitate design and construction management, BMCE developed a bespoke and hybrid digital paradigm which combined 3D modelling and 2D drawings with Asta PowerProject/Synchro 4D BIM simulation as strategic design/project management tools.

It does seem paradoxical that PAS 1192-2:2013 is now fully operational as a construction industry BIM norm without any significant body of evidence that it is fit for purpose or suitably flexible to fit a range of procurement arrangements. Over the next ten years, many and diverse BIM hybrids may evolve and determine that the PAS standard needs to be revisited or even reverse engineered. Manufacturing is already tooling up to apply augmented reality (AR) techniques to process (particularly products in service). In that context, perhaps too much has been written about industry direction of travel from Level 2 to Level 3 BIM, and not enough to address the possibility that BIM's future shape and substance may self-determine as a manifestation of industry need, evidence-based practice and technological advances.

Also, in developing the interests of SMEs and micro organisations, it seems reasonable to argue that the PAS 1192-2:2013 template (by definition) needs to be sufficiently open-ended to be able to accommodate a wide range of paradigms for doing collaborative BIM. Anecdotally, a perception among some SMEs and micro organisations is that engaging with PAS 1192-2:2013 may represent a time hungry on-cost in comparison with current practices. If that is the case, the issue needs to be addressed on a project-by-project basis. Also, the view has been expressed that large organisations may be able to absorb the overheads associated with doing BIM more readily than SMEs and micros.

As with all things BIM, the key to applying PAS 1192-2:2013 is tacit acceptance that digital data is most efficiently managed in a structured way. That concept may not gain universal acceptance in the field, particularly among practitioners. For example, feedback from a regional survey conducted among one professional body group in 2014 suggested that although 55 per cent of respondents were using BIM authoring software, 85 per cent were either not aware of BS1192:2007 or had not used it. For SMEs and micros already pushing (or being pulled) towards collaborative BIM, once the standards set out in BS1192:2007 have been taken on board and implemented, assimilating and applying PAS 1192-2:2013 represents the next logical step on their BIM journey.

There is one significant caveat. In interpreting PAS 1192-2:2013, the language of collaborative BIM may require some translation, even for industry experienced practitioners. Rob Jackson of Bond Bryan Architects produced the informative companion volumes *BIM Dictionary* and *BIM Acronyms* (Jackson 2014a, 2014b) which may assist in familiarisation with BIMspeak. For example, while the term 'workflow' (origins in manufacturing process) is commonly used by BIM practitioners, some might comment that there is something inherently quirky about a sector which has allowed BIM to develop into a distinct subculture with its own arcane language, evangelical leaders and pan-industry tribal following. Having raised that point, any underpinning socio-anthropological nuances and cause/effect consequences are for others to discuss.

14.2 Employer's information requirements

As a key component of the PAS 1192-2:2013 protocol, the term 'EIR' is now common parlance in UK BIM circles and has become embedded into everyday terminology for data management. Historically, a universally accepted starting point for a building project was the commissioning client's brief; there is no escape from the truism that without the client there is no project. PAS 1192-2:2013 does offer guidance that 'EIRs are produced as part of a wider set of documentation for use during project procurement and shall typically be issued as part of the employer's requirements or tender documentation'. Perhaps more fundamental to a BIM project gestation and evolution, the PAS does have a lot to say about the role/scope of the EIR as regards information management. The structural and chronological relationship between the EIR and briefing documents assembled from project inception is less clear.

Does the decision as to whether a building commission should be developed to follow PAS 1192-2:2013 protocols (i.e. as a collaborative BIM project) come from the client, the designer/design-construct team or as an outcome from dialogue between all parties? There is no single answer. Each project will determine its own set of imperatives. As John Eynon argued in response to an article in *BIM+*:

> what clients and professionals should be asking is what information or outputs are needed, and why, [and] what will they be used for and how. This is the point of the EIR. How we get to those required outputs just depends.

SMEs and micro supply chain organisations which have engaged with BIM will be keen to use their acquired knowledge to add value to projects. Expertise in the field may already be embedded into marketing strategies and company blogs. For example, SME Heath Avery Architects in Cheltenham advocate the use of collaborative BIM as an assist for clients and to 'fulfill the design brief with much greater clarity' (Anon 2015a). Bond Bryan Architects blogged that 'visualisation in itself is not "BIM" but as part of a wider approach it still plays a key role in explaining the architectural vision to clients and other stakeholders' (Jackson n.d.).

Central government and many public sector organisations and health agencies and the like have expert client individuals/organisations embedded within procurement teams. These clients may already have the expertise to make timely and informed decisions on the fit of BIM with their strategic and operational objectives. Where a client's portfolio reflects continuity between capital expenditure (CAPEX) and operational expenditure (OPEX), there is logic in thinking and planning long-term from the outset. Expert clients may have established procedures in place for moving information along a data conveyor belt which runs across the whole life cycle of their building estate. In that context, protocols for compartmentalising and managing critical data may have already been embedded in organisational structures pre-BIM.

Paul Wilkinson highlighted the point that not all construction clients have long-term interests in their assets and posed the question 'why would a developer client be interested in asking for BIM deliverables?' (Wilkinson 2013). Ultimately, speculative clients may be primarily focused on selling or letting a new development. On that note, a 2011 RICS survey reported in 2013 suggested that lack of demand from clients may be inhibiting industry-wide adoption of BIM in the UK. With around 8k UK survey recipients, only 35 Quantity Surveying/Building Surveying respondents indicated that some clients intended to use BIM on their projects post-handover and during the occupancy phase.

It is questionable whether the expert client paradigm will apply to smaller projects. Having said that, development of a viable EIR is a key and universal concept embedded within the PAS 1192-2:2013 protocol (Figure 14.1). Karen Alford of the Environment Agency commented:

> During the early stages of our BIM journey we spent some time investigating the activities and the interfaces between ourselves and our supply chain partners. We had candid discussions with some suppliers and it became clear how our approach unwittingly led to ambiguity about our requirements. The introduction of an EIR into the contract documents addresses this ambiguity and is essential for a client who is implementing BIM in their organisation.
>
> (Alford 2015)

SMEs and micros may find that tailoring the Government BIM Task Group EIR template into bespoke variants may best serve business and client needs. In 2013, the BIM4Real forum produced an EIR exemplar for a hypothetical project, and the Consortium of Local Authorities in Wales (CLAW) published comprehensive EIR guidance following a series of workshops intended to assemble a BIM toolkit for local authorities.

Fleshing out and articulating specific data requirements as subsets of building general arrangements and functional spaces needs to underpin client/design team interactions at the early stages of project development. On that basis alone, a specific EIR statement should be included within the project brief. PAS 1192-2:2013 spells out a long list of minimum requirements for EIR content. These condense into three key headers (Figure 14.2):

- technical;
- management; and
- deliverables.

To provide a complete picture for design/project/information management, PAS 1192-2:2013 necessarily needs to be viewed within a broader frame of reference for project documentation. Typically, and for a development following a traditional procurement route, other templates with interdependencies could include BS1192:2007, project-specific versions of RIBA Plan of Work 2013 and NBS BIM Toolkit, plus the NRM 1, 2 and 3 reference documents.

The RIBA Plan of Work 2013 was revamped from previous versions to include a front-end Stage 0 'Strategic Definition' phase. In that context, as the POW 2013 distinguishes between 'Strategic Brief' and 'Project Brief', it is envisaged that the exercise of scoping the EIR (as a key component of Strategic Definition) would be embedded within discussions which reviewed a client's business case for initiating a project.

When framing EIR documentation, it needs to be clarified if design-construct team members will be able to satisfy client and project requirements for collaborative BIM. While PAS 1192-2:2013 does make reference to embedding 'training requirements' in the EIR, it could be regarded as a high-risk strategy to allow consultants to use a live project to develop/test BIM skills, particularly on small projects which might not be subject to a formal and audited prequalification process. In the current competitive economic climate, there is also a risk that implied expertise with the use of digital tools while working in collaborative virtual environments might not be backed up by the ability to deliver.

PAS 1192-2:2013 does also flag up that a project EIR may be written into the tendering process. Plus, embedding the CIC BIM protocol (referred to later) into the mix means that the client/employer is obliged to appoint an information manager. Once the project-specific requirements for doing BIM are plotted out over several reference templates, navigation starts to become more complicated. It is essential to ensure that documents which define and spell out BIM methodology in specific instances read and cross-reference consistently. That's a very old message, which in essence is no different to using paper documentation. The EIR may only be one piece of a much larger matrix of reference documents which feed into project gestation and evolution. Finally, and most significantly, consideration needs to be given not just to templates, rules and inputs, but also to outputs; who needs data, when and in what formats? In that procedural sense, BIM is no different to historical paradigms such as the CPIc conventions which set out templates for paper-based design and project management.

14.3 Writing BIM execution and project implementation plans

A project EIR establishes an enclosing frame of reference for information requirements. That framework is defined using the guidance contained from within PAS 1192-2:2013. The broad intention with the BIM Execution Plan is to plot out a course for information delivery and exchange by mapping out roles, responsibilities and time frames. Drawing from the entirety of the EIR, PAS 1192-2:2013 describes content of the BIM Execution Plan (BEP) as being split over two phases for project delivery: pre and post contract. The Construction Project Information Committee (CPIx) publish downloadable BEP pro formas for both phases. These are useful documents and should cross-reference with the PAS.

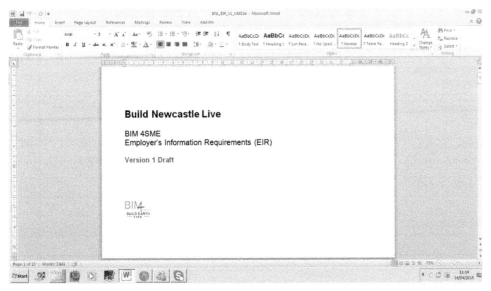

Figure 14.1 Project specific EIR produced by BIM4SME for Build Live competition

According to Federico Negro, Project Director at CASE, the New York-based building information consultancy, the BEP is one of the most important components of any successful BIM project (Negro n.d.). It defines the expected BIM deliverables and guides the co-ordination of the project team. BEPs need to be comprehensive (but succinct) and be tailored to embody project-specific requirements. The CASE methodology begins with a stakeholder workshop to establish project goals and team capabilities. Workshop outcomes lead in to development of a project-specific BEP which defines the people involved in the project, their roles, the process for exchanging, authoring, reviewing and co-ordinating models and the BIM deliverables. At the end of this process, CASE run a project kick-off meeting to take the team through the BEP and ensure that everyone understands the aims, objectives and methods for project delivery. That methodology is generic and can be applied to any size of project.

Some BEP reference documents, for example the AEC(UK) BIM Protocol or New Zealand BIM Handbook, do not distinguish between pre- and post-contract stages with BEP templates and point out that specific projects could either have one or several BEPs depending on clients' strategic and operational requirements. The New Zealand BEP exemplar is particularly useful as it provides the structure for a worked example which is based on a large project, but could be tailored and tweaked to suit smaller commissions.

To be PAS 1192-2:2013 compliant, the pre-contract BEP should define arrangements for producing a Project Implementation Plan (PIP), a Project Information Model (Figure 14.3) a strategy for information delivery, articulate project goals for collaboration/information modelling and define major project milestones along the project programme. The PIP is intended to include an evaluation of 'supply chain capability'. That test of fitness of consultants, subcontractors, suppliers and the like to engage with a BIM process structured under the auspices of PAS 1192-2:2013 would typically be assessed during a prequalification exercise. For example, that evaluation could be undertaken using the PAS 91:2013 template.

To the uninitiated and the BIM novice, this evolving matrix of interdependencies may already seem unwieldy (possibly even unmanageable). There is also a lingering concern that, in

Technical		Management		Deliverables	
Software platforms	√	Standards	√	Defining client's information requirements	√
Data exchange protocols	√	Roles and responsibilities	√	Project team competence assessment	√
Co-ordinate systems	√	Data management	√	Defining project deliverables	√
Model levels of definition/detail	√	Model co-ordination	√	Timing and content of information exchanges	√
Training needs	√	Collaborative process	√		
		Design/project/information management	√		
		Review meeting cycle	√		
		Compliance/audit arrangements	√		
		PIM and AIM management including information transfer	√		

Figure 14.2 Exemplar checklists for developing project specific EIR content

practice, too many potentially constraining parameters may be introduced too early on in the process (for example, before a design brief has been worked up and consultants' appointments have been made). Clearly, having a design team member to assume a BIM lead role from an early stage of project development will assist in the initiation, orchestration, co-ordination and preparation of appropriate documentation.

Production of the post-contract BEP (Figure 14.4) also draws from the EIR as a root source, and sets out a number of operational imperatives under four key headings:

- management;
- documentation;
- standard method and procedure; and
- IT solutions.

It then introduces concepts of the master information delivery plan (MIDP) used to manage information delivery during the project and the task information delivery plan (TIDP) used to convey the responsibility for delivery of each supplier's information. Although SME practitioners may have significant concerns by this stage at the multiplicity of information-related roles which the PAS 1192-2:2013 flags up (variously, lead designer, task team manager, task information manager, interface manager, information originator, project delivery manager, etc.), the document does offer the qualification that for smaller projects these roles may be fulfilled by the same person from within the design and/or construction team.

Ascertaining and defining roles, responsibilities and lines of communication are essential prerequisites for managing digital information in BIM environments. But nothing has

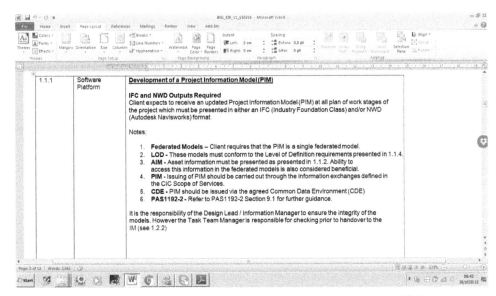

Figure 14.3 Extract from project EIR showing criteria for Project Information Model (PIM)

Management		Documentation and IT Solutions		Standard Method and Procedure	
Roles/responsibilities	√	Project Implementation Plan (PIP)	√	Strategy for defining volumes/spaces	√
Project milestones per programme	√	Agreed processes for collaboration/modelling	√	Project origin and orientation	√
Strategy for content/timing of deliverables	√	Agreed matrix of responsibilities	√	File naming convention	√
Survey data	√	Task Information Delivery Plan. (TIDP) Project team responsibilities for information delivery	√	Layer naming convention	√
Process for approving information	√	Master Information Delivery Plan. (MIDP) When information being prepared and by whom	√	Drawing sheet templates	√
Process for validating data	√	Software/versions	√	Annotation, dimensions, symbols	√
		Data exchange formats	√	Object attribute data	√
		Process/data management systems	√	Classification system/s	√

Figure 14.4 Exemplar checklists for post-contract BIM Execution Plan (BEP) content

changed on that front compared with the use of more conventional methodologies for moving information around design and construction teams. One of the challenges of interpreting PAS 1192-2:2013 generally is that it constructs a dense matrix of interconnected protocols and roles. Jason Fried and David Heinemeier Hansson described the phenomenon as 'thick process' (2010). For small projects, that matrix may require deconstructing and reassembling into a form which embodies key attributes of the PAS, but has less 'operational mass'.

There may be latent dangers in assigning roles specific to information management and also in considering the organisation of digital data out of context from the broader scope of project definition and development. The client's brief, goals and methodologies for achieving these objectives define strategies for information management, not the reverse. To that end, BIM process in its entirety should always be framed within the strategic project brief.

14.4 Determining strategies for CDEs

The Common Data Environment (CDE) is the apotheosis of collaborative Level 2 BIM. The CDE is the virtual repository where digital project information is uploaded, shared and viewed across the design-construct team. An extended function may be to host data which is transferred from pre- to post-handover phases of a building's whole life journey.

The history of shared digital environments had origins in the early days of mainframe computers. When the IBM personal computer came to market in the 1980s, user-purchased software resident on hard drives established a norm which endured for the next 25 years. More recently, the pendulum has swung back and the shift towards centralised online hosting of software applications via 'the cloud' has gained momentum. Commercially available CDEs range from freeware to sophisticated design and project management platforms for multi-user collaboration. For built environment users, the Avanti Programme published construction industry-specific guidance on collaborative working in shared digital environments from 2005. BS1192:2007 provided specific guidance on the structure and operational characteristics of CDEs. That work drove development of the PAS 1192-2:2013 document.

In interpreting the detail of PAS 1192-2:2013, the CDE underpins the process of information delivery, exchange and interactions during design and construction with a collaborative BIM project. Having said that, a CDE is by no means a BIM-specific concept. It can be loosely applied to any digital domain where information is stored, shared and accessible to people with common interests but possibly diverse geographical locations. In that context, social media sites such as Facebook, Instagram and Slack are variants on the CDE theme. They all use virtual environments to link people with data and some host online interactions. Possibly in real time for dialogue. They usually serve as a resource for people to dump data for peers and social networks to view, as well as to retrieve data from them.

The boundaries between digital media being used for social and business dialogue have become blurred. A number of contemporary web-hosted user-sharing applications could be used for social or business purposes (or both). In 2013, Forbes reported that the ubiquitous Dropbox averaged around 175 million users (Rogowsky 2013). Is Dropbox fit for purpose as a CDE for entry-level BIM use? That's an open question. Clearly, for SMEs and micros taking first steps towards collaborative BIM, not only do CDEs need to be useable, they also must be affordable.

Sharing a common characteristic with other generic project management packages, Basecamp® offer a fully functional free-to-use trial version for evaluation. A time-bar limits access, which can be reinstated by converting to a subscription service. Like competitor company Smartsheet, Basecamp® sits between the ubiquitous Dropbox and built environment-specific tools like 4Projects and Asite Adoddle. The Basecamp® online case study charts online interactions between people and process by describing how KEEN Footwear used the CDE to manage the fit-out of its company HQ and flagship store in Seattle, involving 40 people from 10 different companies over an eight-month period (Anon n.d.; Figure 14.5).

Controlling, managing access and ensuring the integrity of data are all key ingredients of a CDE's operational characteristics. The Avanti Programme spelled out the message that a primary aim of the CDE is to reduce or eliminate the checking, revision and reissue cycle of data as a project rolls forward from concept stage towards detail design and construction. In previous iterations of design-construct methodologies, that 'data' would have been embedded on paper drawings, or more recently as digital output via PDF files. For the time being, the fact that 2D output is still necessary for statutory approvals, contract purposes and the like means that "flat" drawings have survived the migration from orthographic to object-oriented environments. Avanti also highlighted some advantages of using CDEs in practice:

- ownership of information remains with the originator although it is shared;
- shared information reduces the time and cost in producing co-ordinated information;
- documents can be generated from combinations of graphic/model files;
- spatial co-ordination should provide 'fit first time' information;
- digital information can be used for estimating/cost/construction planning.

(Avanti 2006)

At the time of writing, best available guidance on the structure and operation of CDE is provided by Mervyn Richards' *Building Information Management: A Standard Framework and Guide to BS 1192* (2010). While BS1192:2007 predated Level 2 BIM definition in the UK, familiarisation with Richards' 2010 guide is an essential precursor to applying PAS 1192-2.

As a first step towards building and practical application of a project-specific CDE, the handbook provides a comprehensive guide to 'Standard Method and Procedure' which sets out various components of the procedural methodology including roles and responsibilities, structural/functional aspects for folder definition and sharing, and standard methods to ensure consistency of file definition and identification. The high-level CDE environment encloses four containers (folders) to host digital files for interdisciplinary working. These are 'Shared', 'Work in Progress (WIP)', 'Published' and 'Archive' and can be applied to any project without qualification on project value, building type and design/construction team discipline mix. The CDE is intended to facilitate:

- co-ordination of project model files (3D + 2D) as they develop;
- production of 2D drawings from 3D models;
- collection/management/dissemination of relevant construction documentation;
- extraction/management of all relevant data (spreadsheets, text files, etc. from data-rich models); and
- application and co-ordination of specification and costing requirements.

The selection of an appropriate host for an online CDE will depend on many factors, not least client-driven imperatives and design and construct team resources. For entry-level engagement with CDEs, small organisations may find it beneficial to trail cloud-based freeware before scoping out more sophisticated packages such as those offered by 4Projects, Asite, Vico, Aconex and a host of other providers. As with BIM authoring software, there is no 'best buy', and precedent suggests that no cloud-hosted digital environment could be ever be truly designated as 'secure'. The range of online design/project management tools available to assist decision-making and smooth information flow needs to be filtered down to match the skills of the design-construct team using them. Also, proprietary CDE tools need to be able to ensure smooth information flow, generate appropriate outputs and offer the facility to provide secure digital environments for cross-disciplinary working.

Figure 14.5 Basecamp® cloud hosted online CDE

14.5 Levels of development/detail

Section 9.8 of PAS 1192-2:2013 provides guidance on 'Levels of Model Definition' (also variously known as 'Level of Development' and 'Level of Detail' and usually abbreviated to LOD). The purpose of the protocol is to facilitate the design-construct team to specify and articulate the content of Building Information Models (BIMs) at various stages in the design and construction process.

These 'models' will typically be generated on a discipline-specific basis depending on the nature of the project and the design-construct team discipline mix. Discipline models authored in data-rich software will generally define 'objects' (elements of building fabric, spaces, rooms, fittings and equipment, etc.) by the geometry of the object (e.g. wall, window, etc. and metadata (alphanumeric textual information attached to the object). The combination of geometry, metadata and the parametric qualities which objects and families of objects usually possess makes for a complex mix. Depending on the circumstances, there may also be direct links between discipline models and the manufacturing process (for example, M&E ductwork models authored in Autodesk Revit MEP could be exported to CAD/CAM fabrication machinery).

For practitioners, contractors, subcontractors and suppliers more familiar with situations where contract documentation is assembled from 2D drawings, paper-based specifications and bills of quantities, the BIM environment may present an intimidating and unfamiliar paradigm. While it could be argued that 'it's just a different way of generating and presenting information', getting to grips with the concept of LOD may present significant challenges, particularly in migrating from 2D orthographic practices which historically separated graphic and alphanumeric into discrete packages and categories for presenting drawn information. The traditional hierarchy for production drawings (general arrangement, assembly, detail) has lost its significance with BIM. Instead, the model represents a densely populated digital database from which information can be extracted as and when required across the whole life of a building (design/develop/construct/use/adapt/reuse/deconstruct/recycle).

Even within the BIMworld, LOD is not a mature concept and may be subject to interpretation. PAS 1192-2:2013 defines phases of project delivery and links maturity of model/s development to each of these stages:

- Brief;
- Concept;
- Design;
- Definition;
- Build and commission;
- Handover and close-out; and
- Operation and in-use.

In 2008, the US architect Jim Bedrick published an article, 'Organizing the Development of a Building Information Model', which introduced the concept of LOD (Bedrick 2011). At that time, the acronym stood for 'Level of Detail', a concept developed by Vico Software for its 'Model Progression Spec'. In 2013, under the auspices of the BIM Forum, the American Institute of Architects (AIA) produced a detailed framework articulating indicative levels of BIM model development. The purpose of the AIA guidance was both to define and illustrate characteristics of model elements representing building systems, assemblies and components at different stages of model evolution, and also to help authors articulate what data their models should provide at different stages of the design process and to facilitate collaborating users to understand the functionality and limitations of models they are receiving.

Key drivers underpinning the AIA's guidance were to help explain the LOD framework and encourage its use so that it became more ubiquitous in everyday BIM practice. The guidance embodies two caveats. First, that discipline models may not all be at the same stage of evolution across the duration of the pre-contract design process. Second, the importance of distinguishing between the terminology 'level of detail' (how much detail is incorporated in an element) and 'level of development' (level of reliable and validated data output embedded in an element). Since, depending on the circumstances, these concepts might be interpreted differently, by different people, the issue of model definition needs to be embedded in the BIM documentation for a project.

As a point of reference, the AIA's LOD protocol is defined at five 'levels' (note that the BIM Forum also defined an intermediate 350 level). These are more specific than the level stages defined within PAS 1192-2:2013 and were underpinned by industry consultation in the USA:

LOD 100: Representation of the model element by indicative geometry only. For example, the architectural 3D concept model of a framed single storey supermarket development could include outlines of structural elements (floor, posts, roof beams) incorporating indicative dimensions based on assumptions.

LOD 200: Graphical representation of the model element as a 3D generic object, or assembly with approximate quantities, size, shape, location and orientation. In the case of the supermarket exemplar, structural grid defined and structural posts (for example) identified as square hollow sections. Non-graphical data such as basic specification information could be attached to the model element. Most discipline specific BIM authoring software would include editable metadata attached to specific model elements as a default.

LOD 300: Representation within the model as a specific system, object or assembly in terms of quantity, size, shape, location and orientation. In the case of a window, for example, LOD 300 could be defined by a manufacturer's proprietary information, for example as a library BIM object imported into the architectural model. Attached metadata would be manufacturer and product specific.

LOD 400: In addition to data sets provided at LOD 300, the model element would typically be represented to include detailing, fabrication, assembly and installation information, either in graphical format, as metadata, or both.

LOD 500: The model element is a validated representation in terms of size, shape, location, quantity and orientation of the real world object incorporating embedded metadata. Referring to the supermarket example, a proprietary heating boiler imported from the manufacturer's BIM object database would be deemed to be at LOD 500 level of development and an instance of a virtual object which would be migrated from design/construct to facilities management phases of a building's whole life

Depending on the model element, there might be little practical difference in representation between LOD 200–400. For example, the hollow steel column referred to may have been significantly defined at LOD 200, while the boiler referred to could remain as an abstraction until manufacturer specific information had been substituted (typically as BIM objects) at LOD 400–500. There is no hard and fast rule; at this point in time, the concept of LOD is open to interpretation within the broad PAS 1192-2:2013 framework. Whatever project-specific templates may be applied to the concept of LOD, the key driver is that protocols are agreed and defined in the BIM documentation, particularly with regard to required outputs at each defined milestone throughout the BIM process.

For the time being, the process of abstracting design intent through an evolving process of sketch design, detail design and production drawings remains embedded in built environment culture. Small projects may work with supporting textual information included on 2D drawings. Larger commissions may require separate written specifications and bills of quantities. Implementation of the BIM process may be hugely challenging for small organisation to get to grips with and take on board, particularly with a collaborative project developed under the PAS 1192-2:2013 umbrella. Understanding and embracing the concept of LOD is a significant factor in making the transition from contemporary towards BIM paradigms for project development and documentation.

14.6 The fit and use of industry organisational standards: the CIC protocols

Consider a hypothetical scenario to develop a single storey rural visitor/interpretation centre with a gross floor area of 130 m² and a budget of £500,000 using a traditional procurement route. A lead designer was appointed by the client, a strategic/project brief prepared and a concept design assembled. The project will be awarded following competitive single-stage tendering process. The lead designer has previous experience of collaborative BIM and has agreed to assume a co-ordinating role for design and information management at the client's request. The client has made separate appointments for structural engineering, quantity surveying and services engineering. The RIBA Plan of Work 2013 has been adopted as a management process model.

The visitor centre is the third in a portfolio of similar developments which the client will self-manage after handover. It is envisaged that data embedded in discipline-specific BIM models during design/construct will be filtered and used by the client for post-occupancy planned/reactive maintenance. A BIM process has been agreed based on the use of federated models for project development; i.e., each discipline will produce 'own models' and these will be combined in a viewing environment for validation of spatial arrangement and geometry. All design disciplines are using variants of data-rich BIM authoring packages produced by the same software house, and data exchange will be handled using native files (the file type used by originating software). Outputs from each design discipline will be 2D PDF files. File management will be co-ordinated by the lead designer. If the discipline-specific BIM authoring packages use different native file types, the IFC file standard can be used as a common language for file transfer/interaction (see Bond Bryan Architects, AMRC project Rotherham for case study discussion).

It is assumed that BS1192:2007 and PAS 1192-2:2013 reference standards will be applied to the project, and the design lead will co-ordinate production of the EIR, BEP and ancillary planning documentation for information management. The CIC BIM Protocol (authored by Government Task Group and Construction Industry Council and endorsed by most built environment professional bodies) may also be in the frame of reference. The CIC pro forma is intended to act as a supplementary legal agreement for incorporation into consultant appointments and the construction contract. The overarching intention with the protocol was to formalise support for collaborative working in a Level 2 BIM environment.

The CIC template cements contractual relationships between the client and suppliers of services (for example, consultancies) and goods when using a BIM methodology for information management. In the lead-in period to confirming contract arrangements, suppliers' capabilities may be tested using the PAS 91:2013 prequalification gateway. The CIC BIM Protocol was also designed to be used by construction and contractor clients, including for managing the work of subcontractors.

Making reference to the exemplar project, each of the four directly appointed consultants would use the same completed pro forma plus appendices as an addendum to their terms of agreement with the client. The CIC protocol's authors also intended that, for example, following a tender process, the successful contractor could utilise the same template to manage the work of subcontractors. There is sufficient flexibility embedded in the CIC BIM document to be adaptable to suit a range of procurement methodologies, including design and build.

Simon Lewis (2013) noted that the CIC BIM template was structured in a tabular format first established by the AIA BIM Protocol Exhibit 202. That paradigm required the production of what is referred to in the protocol as a Model Production and Delivery Table (MPDT), which specifies relevant levels of detail (LODs) and stage definitions showing models required at each defined stage of project evolution (the BIM Task Group website hosts a MPDT template for reference). Application of that procedure links information exchange milestones (data drops) with relevant model originators and LODs across each stage of whatever process map is adopted by the design-construct team. (In the case of the exemplar project, RIBA Plan of Work 2013 is being used.) The purpose of the procedural mapping is intended to ensure that the design-construct team formally buy into designation of roles and responsibilities in driving the BIM process forward.

A specific requirement of the CIC BIM template is that an Information Manager is nominated and appointed by the client with the qualification that 'the Information Manager has no design related duties. Clash detection and model co-ordination activities associated with a BIM Co-ordinator remain the responsibility of the design lead.' The Information Manager's key role is identified by CIC as managing the processes and procedures for information exchanges, plus:

- initiating and implementing the Project Information Plan and Asset Information Plan (as applicable);
- co-ordinating the preparation of project outputs (for example, data drops); and
- implementing the BIM Protocol, including the updating of the MPDT.

The diversity of roles articulated across the scope of BS1192:2007, PAS 1192-2:2013 and the CIC BIM Protocol may cause some confusion. For a small project as outlined in the exemplar, it is difficult to see beyond the situation where a nominated team member would assume overall responsibility for co-ordinating design/information management. The issue of change management, for example, is generic, not BIM-specific, and robust procedures for BIM management need to be nested within strategic design/project management protocols which are transparent, workable and not so time-consuming to implement that they load up project on-costs.

The CIC BIM Protocol embodies a procedural approach. It works by determining project-specific requirements, adopting an agreed set of procedures and following these consistently through out the design-construct process. Whether the perceived open-endedness of the document is perceived as a strength or weakness needs to be market tested to provide evidence-based feedback. Because of its contractual nature, the document does flag up several nuances of interpretation which have not been fully tested at the time of writing. Principally, these relate to ownership of data (intellectual property) and possible liabilities associated with electronic data exchange between model originators (author/s) and receivers/users across the organisations which comprise the project teams.

The CIC BIM pro forma is a case in point which reinforces the message that the key to efficient and reliable management of digital information is a structured approach to file organisation, attention to detail and rigorous cross-referencing between elements of BIM documentation to ensure consistency of interpretation and consistent application at project level. Like PAS 1192-2:2013, the CIC BIM Protocol will gain added credibility when it has been road-tested through practice and feedback has been incorporated into the evolutionary process embedded in uptake and use of Level 2 BIM documentation. Weaving layers of complexity into the configuration and interpretation of BIM standards will not necessarily ease the passage of UK construction towards meeting desired Government aspirations and outcomes. Perhaps sometime in the future a viable formula for an holistic BIM protocol attractive to SMEs and micros will emerge from the plethora of documentation currently in the frame. Once the NBS BIM Toolkit has been fully tested, shaken down and refined, it may (partly or wholly) deliver on that point.

Bibliography

Alford, K. (2015) 'BIM for Clients – an EIR is Key', *Adjacent Digital Politics Ltd*, 2 February, [online]. https://www.adjacentgovernment.co.uk/housing-building-construction-planning-news/bim-clients-eir-key-2/ [accessed 2 February 2015].

Anon (2012a) *AEC (UK) BIM Protocol Project BIM Execution Plan*, AEC (UK), [online]. https://aecuk.files.wordpress.com/2012/09/aecukbimprotocol-bimexecutionplan-v2-0.pdf [accessed 2 February 2015].

Anon (2012b) 'Collaborative Design Makes a Big Impact for Nuclear Advanced Manufacturing Research Centre', Graphisoft UK Ltd, [online]. http://download.graphisoft.com/ftp/marketing/case_studies/NAMRC_GRAPHISOFT_CaseStudy.pdf [accessed 2 February 2015].

Anon (2013a) 'Employers Information Requirements', BIM 4REAL, [online]. www.bim4real.co.uk/Downloads/c1/eir.pdf [accessed 30 January 2015].

Anon (2013b) *Employers Information Requirements Core Content and Guidance Notes*, Version 07, BIM Task Group, [online]. www.bimtaskgroup.org/wp-content/uploads/2013/04/Employers-Information-Requirements-Core-Content-and-Guidance.pdf [accessed 30 January 2015].

Anon (2013c) 'PAS 1192-2 – Overview', HM Government Department for Business Information and Skills, [online]. www.bimtaskgroup.org/PAS11922-overview/ [accessed 30 January 2015].

Anon (2014a) *New Zealand BIM Handbook*, The Building and Construction Productivity Partnership Ltd, [online]. www.buildingvalue.co.nz/sites/default/files/New-Zealand-BIM-Handbook.pdf [accessed 2 February 2015].

Anon (2014b) 'Software as a Service – Understanding Your Options', Mason Advisory, [online]. www.masonadvisory.com/white-paper/software-service-understanding-your-options [accessed 3 February 2015].

Anon (2015a) 'Building Information Modelling', Heath Avery Architects, [online]. www.heath-avery.co.uk/bim/ [accessed 30 January 2015].

Anon (n.d.a) *The CLAW All Wales Toolkit*, Consortium of Local Authorities in Wales, [online]. www.cewales.org.uk/cew/wp-content/uploads/4.-EIR-no-boxes.pdf [accessed 30 January 2015].

Anon (n.d.b) 'Here's How KEEN Footwear Used Basecamp to Help Build their Flagship Store in Portland', Basecamp, [online]. https://basecamp.com/tour [accessed 24 March 2015].

Avanti (2006) Project Information Management: A Standard Method & Procedure. Avanti Toolkit 2 Version 2.0, January, [online]. www.avanti-construction.com/docs/Avanti%20Toolkit%202%20(V2.0)%20SMP%20-%20Jan%202006.pdf [accessed 19 August 2015].

Beale & Co. (2013) *Building Information Model (BIM) Protocol*, 1st edn, Construction Industry Council & BIM Task Group.

Bedrick, J. (2011) Organizing the Development of a Building Information Model 2008, 2011, 26 April 2011, *AEC Bytes*, [online]. www.aecbytes.com/feature/2008/MPSforBIM.html [accessed 9 February 2015].

(2013) 'A Level of Development Specification for BIM Processes', *AECbytes*, 16 May, [online]. www.aecbytes.com/viewpoint/2013/issue_68.html [accessed 9 February 2015].

BICS (2011) *RICS 2011 Building Information Modelling Survey Report*, Building Cost Information Service, Royal Institution of Chartered Surveyors, [online]. www.rics.org/Global/RICS_2011_BIM_Survey_Report.pdf [accessed 28 January 2015].

BSI (2013) *PAS 1192-2:2013 Specification for Information Management for the Capital/Delivery Phase of Construction Projects Using Building Information Modelling*, BSI Group, [online]. http://shop.bsigroup.com/Navigate-by/PAS/PAS-1192-22013/ [accessed 28 January 2015].

CIC (2013) *Building Information Model (BIM) Protocol*, 1st edn, Construction Industry Council.

Cole, S. (2013) 'VW Shows Off iOS Augmented Reality Repair App for XL1 Concept Car', *appleinsider*, 30 September, [online]. http://appleinsider.com/articles/13/09/30/vw-shows-off-ios-augmented-reality-repair-app-for-xl1-concept-car [accessed 4 February 2015].

Fried, J. and Heinemeier Hansson, D. (2010) *Rework, Change the Way You Work Forever*. London: Vermilion.

Hayman, A. (2013) 'Clients Impeding BIM Take-up', *Building*, 3 May 2013, [online]. www.building.co.uk/404Handler.aspx?aspxerrorpath=/news/clients-impeding-bim-take-up/5054254.article [accessed 28 January 2015].

Jackson, R. (2014a) *BIM Dictionary*, Bond Bryan Architects (with Scott McKinnell, Tim Platts, Karl Redmond, Duncan Reed and Paul Woddy), [online]. http://bimblog.bondbryan.com/wp-content/uploads/2014/10/30004-BIM-dictionary.pdf [accessed 28 January 2015].

Jackson, R. (n. d.) 'Vision to reality: Bradford College case study' *BIM Blog* (Bond Bryan Architects) [online] http://bimblog.bondbryan.com/vision-to-reality-bradford-college-case-study/ [accessed 24 September 2015].

(2014b) *BIM Acronyms*, Bond Bryan Architects, [online]. http://bimblog.bondbryan.com/wp-content/uploads/2014/10/30001-BIM-acronyms.pdf [accessed 28 January 2015].

Kliskey, T. and Tavendale, J. (2013) 'V&A at Dundee Design and Engineering', (unpublished lecture). Institution of Mechanical Engineers, 21 November, [online]. http://nearyou.imeche.org/near-you/UK/Scottish-Region/Tayside–North-Fife-Area/event-detail?id=8620 [accessed 28 January 2015].

Lewis, S. (2013) 'The CIC BIM Protocol: Some First Thoughts', *Building Information Modelling*, NBS, [online]. www.thenbs.com/topics/bim/articles/cicBIMProtocolSomeFirstThoughts.asp [accessed 9 February 2015].

McCrae, A and King, A. (2005) *An Introduction to Avanti*, Constructing Excellence in the Built Environment, [online]. www.avanti-construction.com/docs/Avanti%20Report%20No%201%20-%20Introduction%20to%20Avanti%20-%20March%202005.pdf [accessed 3 February 2015].

Maqbool, A. (2015) 'BIM Bytes: To BIM or Not to BIM?, That Is the Question for Designers', *BIM+*, 21 January, [online]. http://bim.construction-manager.co.uk/management/question-designers/ [accessed 9 February 2015].

Negro, F. (n.d.) 'BIM Execution Planning' CASE, [online]. www.case-inc.com/service/bim-execution-planning [accessed 24 March 2015].

Richards, M. (2010) *Building Information Management: A Standard Framework and Guide to BS 1192*, BSI Group.

Richards, M., Churcher, D., Shillock, P. and Throssell, D. (2013) 'CPIx Bim Execution Plan', *CPIx Online*, Construction Project Information Committee, [online]. www.cpic.org.uk/cpix/cpix-bim-execution-plan/ [accessed 19 February 2015].

Rogowsky, M. (2013) 'Dropbox Is Doing Great But Maybe Not As Great As We Believed', *Forbes*, 19 November, [online]. www.forbes.com/sites/markrogowsky/2013/11/19/dropbox-makes-hundreds-of-millions-so-why-is-it-only-asking-for-an-8b-price/ [accessed 4 February 2015].

Shah, R. and Bebbington, A. (2013) '4D BIM Case Study: New South Glasgow Hospitals (NSGH)', Brookfield Multiplex Europe, [online]. www.brookfieldmultiplex.com/projects/europe/uk_scotland/construction/health/under_construction/south_glasgow_university_hospital_and_royal_hospital_for_sick_children/ [accessed 28 January 2015].

Wilkinson, P. (2013) BIM: 'Why Should Clients Be Bothered?', *Extranet Revolution*, 23 December, [online]. http://extranetevolution.com/2013/12/bim-why-should-clients-be-bothered/ [accessed 29 January 2015].

15 Project-specific process templates, definition and delivery

15.1 RIBA Plan of Work 2013, NRM and the NBS BIM Toolkit

One of the many challenges facing SMEs and micros in getting to grips with BIM is applying formal industry templates for information management to the processes of responding to client requirements. Every project, no matter how small, needs to be mapped over time from the outset so that inputs/outputs in relation to client expectations can be tracked and managed between design, construction, handover and post-occupancy phases where applicable. How effectively the 'eight pack' (see Section 13, Figure 13.1) suite of BIM protcols for Government-procured work will transpose to smaller projects without dilution and loss of purpose remains to be fully tested. Also, the PAS suite is supplementary to industry-standard process models already in the frame of reference: RIBA Plan of Work 2013, NRM 1,2,3 and, more recently, the NBS BIM Toolkit.

In considering a slice through the membership of a typical built environment supply chain, possibly some constituents are more sensitive to the need to engage with BIM than others. For example, with the range and volume of proprietary BIM objects now being catalogued and held in repositories like bimstore.co.uk®, BIMobject® and SpecifiedBy, manufacturers and suppliers who are heavily dependent on market forces to sell on their products may be more sensitive of the need to be BIM aware than other construction professionals. The BIM4 Manufacturing (BIM4M2) group is still relatively new, having only been set up in 2014 following a request by Construction Products Association for product manufacturers to have representation on the Government BIM Task Groups.

Practitioners are happy to talk about the software they are using and discuss perceived benefits or challenges with BIM, but less forthcoming when it comes to unpicking to what extent Level 2 protocols are being applied to process; in particular to collaborative working. The PAS 1192 suite is only one of a number of process protocols which may fall within the frame of reference when making buildings. It is essential these 'templates' are considered together and harmonised where necessary to facilitate the smooth flow of information and avoid possible conflicts of purpose and detail. The process of co-ordination between diverse information sources is not a function exclusive to the digitalisation of construction.

Anecdotal evidence from CIAT practitioners suggests that small organisations may be reluctant to engage with Level 2 BIM processes, particularly if the 'set-up' phase is perceived as being time hungry, potentially adding another layer of administrative process and cutting further into already slender financial margins. On the other hand, SMEs and micros aspiring to form partnerships with clients and supply chains where BIM is already established as a methodology may need to upskill in order to remain competitive in the marketplace.

The previous Section discussed PAS 1192-2:2013 as an industry template for the application of Level 2 BIM in practice. However, managing digital data is only one of many necessary iterative actions/interactions which contribute to collaborative teamworking. The key function with information management is to inform decision-making and practical actions in responding to commissioning client expectations. In that context, the most enduring and overarching construction process template since its first launch in 1963 and publication the following year has been the RIBA Plan of Work.

The RIBA Plan of Work 2013 divides project delivery into eight work stages (0–7), each with clearly defined end points or project milestones. The protocol also details typical tasks and outputs required during each phase of project evolution. Online guidance is supplemented by an online tool which can be used to produce bespoke and project-specific Plan of Work templates. A BIM overlay was published in 2012 as an outcome from an earlier review of the Outline Plan of Work 2007.

While the Plan of Work is used to map out and monitor design process, the standards which cover cost planning and financial management are the New Rules of Measurement (NRM) 1,2,3 published by the Royal Institute of Chartered Surveyors. The NRM suite provides an industry standard for estimating, cost planning, procurement and whole life costing in the UK. NRM-1 was first published in February 2009 and provides guidance on the quantification of building works in order to prepare order of cost estimates and cost plans as well as approximate estimates. It also includes procedural information about quantifying wider costs, such as preliminaries, overheads and profit, risk allowances and inflation.

NRM2 was published in 2012 and became operative in 2013, replacing the Standard Method of Measurement, seventh edition (SMM7). NRM2 established detailed measurement rules for bills of quantities preparation, including guidance on the content, structure and format of bills of quantities (BOQs). NRM3 was published in March 2014. It allowed the quantification and description of maintenance works, including order of cost estimates and general cost plans. It also provided guidance on procurement and cost control.

Stuart Earl, chair of the RICS Measurement Initiative and a director at Gleeds, argued that the NRM suite was developed to ensure that at any point in a building's life there will be a set of consistent rules for measuring and capturing cost data. As BIM is primarily intended to address data use over the whole life of buildings, compared with the previous Standard Method of Measurement, there is shift of emphasis from task-oriented activities towards an more holistic perspective for data management. It is also important to take on board NRM's purpose as a toolkit for cost management, not just a set of rules for quantifying, scheduling and costing building work. Earl summarised content of the NRM suite as follows:

- NRM1: order of cost estimating and cost planning for capital building works. This underpins how we budget and design our buildings.
- NRM2: detailed measurement for building works. This is a supporting set of detailed measurement rules enabling work to be bought either through bills of quantities or schedules of rates for capital or maintenance projects.
- NRM3: order of cost estimating and cost planning for building maintenance works. This enables the measurement of capital cost plans to be integrated with maintenance and life cycle replacement works.

(Earl 2012)

The aspiration is that clients will be encouraged to use the tools within the wider NRM suite and embed them into their delivery systems. For example, from a cost planning perspective, clearly defining cost-related outcomes at the conclusion of each RIBA Plan of Work

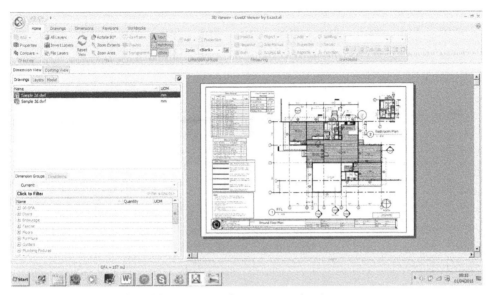

Figure 15.1 Quantities take-off from 2D general arrangement drawing

stage should be a transparent method of ensuring that clients have access to best information at all times. Also, if unforeseen circumstances arise and have cost planning implications, these can be dealt with quickly, efficiently and in collaborative environments. As Stuart Earl also noted, these 'rules' apply throughout supply chain structures, not just to the QS. During the estimating and cost planning stages, designers can be better informed as to what the wider team is planning to achieve and how the rules can be applied consistently to facilitate teamworking in attaining client objectives.

Cost take-offs from 2D drawings tend to follow established hierarchical formats of general arrangement (Figure 15.1), assembly and detail. As Louise Sabol pointed out (2008), in the absence of definitive standards, the levels and densities of data (geometric and metadata) encapsulated within BIM objects at different stages of project development are currently a judgement area. For cost planning purposes there is a pressing need to develop standards for the representation of BIM model building elements (objects, assemblies, systems) at various stages of development between RIBA Plan of Work Stages 2–4. 'Elements can be modeled in generic fashion (e.g. *Flush Door*) or to a well-defined level of detail (e.g. 5-Ply Particle Core WDMA Extra Heavy Duty PC-5 Flush Door)'. At any point in the design process, there is a balance to be struck between creating 3D data models which will be detailed enough to support QS take-offs (Figure 15.2), and generating overly complex models which are difficult to manipulate, co-ordinate and regularly update.

Sabol also noted that objects and assemblies in a BIM environment can be encoded with data which facilitates cost planning (Figure 15.3). Additional information can be affiliated with objects in external databases, for example as might be useful to feed post-occupancy activities. Costing software applications are able to harvest information from BIM models and constituent objects in various ways. For example, by relating to object definitions within the model, or using a unique identifier to link objects to more detailed information stored externally from the BIM application in a database such as Microsoft Access or Oracle.

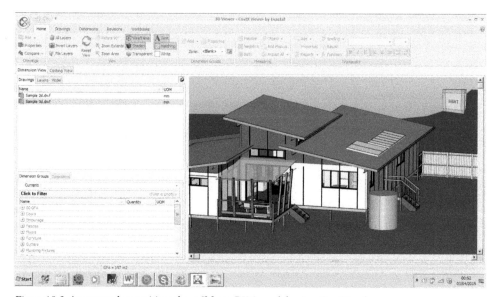

Figure 15.2 Automated quantities take-off from BIM model using CostX software

Figure 15.3 Proprietary window as a BIM model encoded with geometry and attached metadata sourced from the BIMobject® free-to-use online catalogue of construction objects

As with manual take-offs, it is the currency and veracity of the data which are important. Dick Barker of Laing O'Rourke commented that:

> due to the speed of response created by auto-measurement we are able to influence and inform the design as it develops with the knowledge of quantum and cost, not just measure and cost the design, which is often too late.

(Barker 2011)

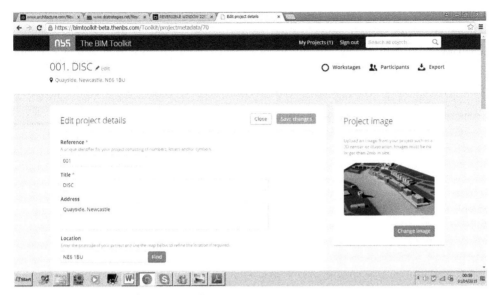

Figure 15.4 NBS BIM Toolkit project editing screen

The RIBA Plan of Work and Standard Method of Measurement (as superseded by the NRM suite) are long-established industry process models. In 2014, following a national competition, NBS was awarded a Government contract, from the BIM Task Group in association with Innovate UK, to undertake the development of a digital BIM toolkit. The NBS digital tool was intended to be the final piece of UK Government's Level 2 BIM 'jigsaw' to provide a comprehensive template for setting up a Level 2 BIM project and mapping project information across a range of process and product models, including the RIBA Plan of Work and classification protocols.

Launched in April 2015, the first publicly available beta version of the NBS BIM Toolkit offered a cloud-hosted free-to-use web portal (Figure 15.4) which provided step-by-step support for defining, managing and validating information development and delivery at each stage of the project life cycle. The digital tool is structured to track RIBA Plan of Work 2013 work stages and was designed as a repository to host open, shareable and verifiable asset information (see also Section 16 in relation to classification).

As the project information manager inputs data, a key attribute of the NBS software is that it generates a bespoke and editable database which links people, tasks and information with time frames for project delivery. The NBS tool was designed to be BIM rather than designer led and, as information can be exported in MS Word and Excel formats, it can be used to populate BIM-supporting documentation, including pro forma EIRs and BEPs. That attribute in itself implies a level of interoperability between constituent parts of BIM documentation.

The BIM Toolkit also organises component, assembly and building related data using a pan-industry evolution version of Uniclass which is aligned to the draft ISO/DIS 12006-2 international standard. The tool was designed around a unified digital structure which provided mapping and guidance so objects could be configured at a project level to have the correct digital classifications as required. Supplemented by step-by-step support for the definition, management and validation of BIM data within project-specific environments.

Templates are under development to set out guidance for Levels of Detail (LODs) and Levels of Information (LOIs) for construction objects. Initially, these will be spaces, systems and products for architecture, building services, structural engineering, landscape design and civil engineering. These are available in both IFC and MS Excel formats and form the protocols which all project teams can use to define information exchanges at a specific project stage.

Rob Jackson of Bond Bryan Architects commented:

> I just want to say that the toolkit is a massive positive for the industry. The team who have put this together have only had 6 months (including a Christmas period) and what has been achieved in the time is impressive. It's a great start and really pushes the Level 2 agenda forward.
>
> (Jackson 2015)

Four industry templates which support BIM documentation have been discussed. Individually, these models fall into two categories; the PAS 1192 suite and NBS BIM Toolkit are BIM specific protocols. The RIBA Plan of Work 2013 and NRM 1,2,3 are industry 'standards' which may include BIM projects. They are all process defining, recording and management tools which can be applied to BIM projects at every level, ranging from small-scale, one-off buildings to large complex developments.

Perhaps the trick with implementation of the templates is less about their value as independent templates and more about how they can effectively be merged to form the core of an holistic strategy for project evolution and delivery to meet client expectations. On larger projects that task is likely to be fulfilled by dedicated design and information managers. On smaller projects, the roles will invariably be combined. There is no hierarchy implied with the process models discussed. In the case of a traditional procurement route, the standards can be regarded as mutually interdependent constituents of a formula for efficient project delivery.

15.2 Strategic Definition and Briefing (Stages 0–1)

Published in 2012, RIBA's BIM Overlay offered a template which could be superimposed over the 2007 Outline Plan of Work (updated 2008) based on a set of defined Work Stages A–M. These have been superseded by Plan of Work 2013 (POW). Supplementary to the POW, the RIBA Plan of Work Toolbox is a downloadable spreadsheet, in Microsoft Excel format, containing customisable tables allowing easy creation of the Project Roles Table, Design Responsibility Matrix and Multidisciplinary Schedules of Services. Similarly, the NBS BIM Toolkit can be populated with project details as they become available.

At this point, a conceptual study to develop a Newcastle Digital Innovation and Showcase Centre (DISC) on a Tyneside frontage forming part of the Ouseburn Regeneration Area is referred to. The project acted as an exemplar to illustrate key aspects of embedding and managing digital information using the Plan of Work 2013 as a process model. The 1.7 hectare site for the project was formed by a narrow strip of south-west facing riverside frontage currently used as surface car parking. The client's vision was to create a new unified building facility where digital and technological innovations could be inspired and nurtured; a destination where new ideas could be turned into new businesses, products and jobs.

A multidisciplinary project team was drawn from SMEs and micro organisations, and an expert client with knowledge of the PAS 1192-2 process was seconded from a local authority to provide guidance on BIM-related information requirements and supporting documentation (EIR, BEP, Soft Landings Strategy). A feasibility exercise was assembled during an intensive 48-hour period of online collaborative teamworking from various locations within the

UK. Key strands of the decision-making process and sample stage outputs are highlighted below. Notes on Stages 4–6 are hypothetical projections drawn from data generated during the exercise.

Stage 0: Strategic Definition

- confirmation that the client's business case for the development had been made and that funding was secured;
- indicative project programme established;
- early engagement of design team consultancies: architect, structural engineer, energy consultant/services engineer, quantity surveyor and information manager all actioned as separate appointments. Principal Designer role allocated (CDM 2015);
- review of design team competencies, software capabilities and establish training requirements for incorporation into employer's information requirements (EIRs) and BIM execution plan (BEP);
- preliminary assessment of client's information requirements, including security of online and cloud-hosted data;
- client requirement for a 'zero carbon' strategy identified in line with the goals of the UK Green Building Council (UKGBC) as a whole life project deliverable;
- BREEAM 'Excellent' rating embedded as a performance requirement;
- PAS 1192-2 standard incorporated as overarching standard for information management. RIBA Plan of Work 2013, NRM and NBS BIM Toolkit adopted as process tools;
- soft landings methodology embedded (including POE to monitor energy performance in use) based on the BSRIA 5-stage process.

Stage 1: Preparation and Brief

- client brief confirmed (including whole life information requirements). Health and safety strategy (CDM 2015) embedded;
- project programme drafted and reviewed;
- consultant contractual arrangements confirmed. CIC BIM protocol adopted;
- procurement strategy determined; JCT single-stage tender with bills of quantities;
- BIM kick-off meeting. Scope of BIM modelling environment and BIM inputs/outputs agreed. Relationship of Project and Asset Information Models confirmed. Draft EIR and BEP produced. Software, file formats, exchange mechanisms and cyber-security arrangements agreed;
- terrestrial high-definition surveying (HDS) laser scan site survey carried out as direct appointment with specialist (Figure 15.5);
- Common Data Environment (CDE) structured, set up and compliance-checked with BS1192:2007 standards for information management. Online project collaboration tool and NBS BIM Toolkit set-up and trialled by information manager;
- site investigation desktop and site borehole surveys commissioned by structural engineer;
- options analysis study prepared by design team; preliminary consultations with planning authority to establish best fit with client requirements and area regeneration strategy;

Figure 15.5 High-definition laser scan of Newcastle quayside site for DISC project by FARO UK, saved as Autodesk Recap native file

- draft carbon management and soft landings strategies. These are subject to an ongoing development/validation with milestones embedded in the Plan of Work 2013 as Sustainability Checkpoints.

Information exchange at Stage 1 (data drop 1) includes:

Client brief, project programme, options analysis study, digitised site survey. Draft documentation for information management strategy (EIR and BEP).
Sustainability checkpoint 1 (carbon and soft landings strategies).

Note on Stages 0–1

To ensure projects are effectively managed and validated as they develop, data which reflects design team work-in-progress is submitted to the client at defined milestones. PAS 1192-2 describes these information exchange points as 'data drops'. The milestones and formats for data exchange are defined in the project EIR and BEP.

The RIBA Plan of Work 2013 sets out six information exchange points at the conclusion of each of Stages 1–6, but these may vary depending on the type of project, client requirements and procurement route adopted. Similarly while the exact nature of the validation process may be project specific, the information manager takes overall responsibility of ensuring that digital project data is consistent, fully co-ordinated and current. That role is referred to in the CIC Building Information Model (BIM) Protocol.

The generation/validation of discipline specific data, followed by validation of shared data by the information manager is cyclical and a continuous process across Plan of Work 2013 stages. Note there are detail differences in 'work stages' as defined by PAS 1192 and RIBA Plan of Work 2013. Extracting and manipulating COBie data did not form part of the exercise.

15.3 Design development, cost planning, technical design (Stages 2–4)

By the conclusion of Stage 1, the client was financially committed to developing the project, a brief had been developed, a design team had been commissioned and a procurement methodology and programme had been established. Key protocols/actions for information and design management had been put in place. The options analysis had been reviewed and updated, and a preferred design concept tabled for development by the multidisciplinary team. The results of the site investigation desk study and site borehole survey were available.

Protocols were established for agenda/frequency of online team meetings. Software requirements and types were agreed. Methodologies for model generation, federation, data exchange and outputs were reviewed to allow the EIR and BEP documentation to be finalised and signed off by the client. An outline strategy for incorporating significant prefabrication and off-site manufacturing elements into the methodology for project delivery was embedded into design team thinking.

Stage 2: Concept Design

- carbon management and soft landings strategies reviewed and confirmed;
- model and change management procedures agreed;
- project EIR and BEP finalised, confirmed with client and embedded in CDE;
- design disciplines prepared concept proposals for architectural, structural and services input;
- concept proposals and project programme reviewed;
- health and safety strategy per Construction (Design and Management) Regulations 2015 embedded;
- preliminary cost plan prepared from concept scheme and site model;
- architectural, structural and services concept BIM models in preparation;
- validation of 'own concept models' by individual disciplines;
- initial model sharing with design team for validation, strategic analysis and options appraisal;
- protocols for file sharing and exchange between disciplines test-driven and appraised;
- model data extracted by environmental designer used for environmental performance and area analysis;
- protocols for defining key model spaces and elements and staged LOD agreed;
- client provided with access to BIM data for feedback using proprietary model viewer.

Information exchange at Stage 2 (data drop 2) includes:

Finalised EIR, BEP, carbon management and soft landings strategies. Draft health and safety plan. Concept scheme proposals. Updated project programme. Preliminary cost plan.
Sustainability checkpoint 2 (carbon and soft landings strategies).

Stage 3: Developed Design

- scheme design of architectural, structural and services models;
- individual discipline validation of 'own models';
- discipline models federated in shared environment using proprietary software and dimensional co-ordination check (clash detection) undertaken by information manager;
- outline specification information embedded into discipline models;
- predicted energy performance prototyped using specialist environmental design software data sharing and integration for design co-ordination;
- soft landings strategy reviewed against client and user interactions and updated;
- preliminary run of project planning requirements (4D) using specialist software;
- review of models to determine generic and bespoke components for appraisal by potential specialist subcontractors and suppliers;
- client approval for developed design;
- saved orthographic views extracted from architectural model for planning submission.

Information exchange at Stage 3 (data drop 3) includes:

Finalised project programme and health and safety plan. Updated cost plan. Preliminary energy appraisal report. Orthographic views extracted from architectural model as design drawings for planning submission to local authority and client records. Sustainability checkpoint 3 (carbon and soft landings strategies).

Stage 4: Technical Design

- detailed modelling integration and analysis by individual disciplines;
- dimensional co-ordination inputs/outputs updated by information manager;
- review and update of specification requirements;
- data extraction from BIM models for bill of quantities preparation;
- creation of detailed model elements for design disciplines;
- all disciplines check of technical requirements for building control and BREEAM;
- final run of specialist energy appraisal software to identify compliance with BREEAM performance standard for energy use;
- saved orthographic views exported from discipline models for building control submission;
- pre-contract appraisal procedures initiated including pre-qualification process for suppliers using PAS 91:2013;
- cost plan reviewed and updated against bill of quantities and finalised specification requirements;
- model outputs updated for tender action and level of access of tenderers to models confirmed;
- client approvals formalised following project sign-off by building control;
- finalised documentation prepared and tender action initiated;
- tender review and cost plan updated.

Information exchange at Stage 4 (data drop 4) includes:

BREEAM assessment. Building control submission including energy appraisal. Bill of quantities. Tender documentation including appropriate model outputs; drawings, schedules. Health and safety, energy appraisal and soft landings reports.
Sustainability checkpoint 4 (carbon and soft landings strategies).

Note on Stages 2–4

As discipline information is developed and logged to the CDE, the information manager's role is critical in maintaining consistency and currency of shared data before it is 'published', i.e. made available to third parties external to the design team during the pre-contract phase of project evolution.

Protocols for the validation of digital data need to be embedded both in discipline-specific and shared (for example, federated) model environments as agreed and recorded in the project BEP.

Depending on the levels of collaboration embedded in the project, some activities (for example, development of health and safety protocols) may be evolving outwith the BIM environment. These tasks also need to be managed and may not fall within the information manager's remit. Whichever arrangements for information management are in place, either a design team member or client nominated specialist needs to be responsible for overall co-ordination of the project.

When using a design and build procurement methodology, early contractor/sub-contractor involvement offers more scope for concurrent workflows than a traditional procurement route.

15.4 Construction and Handover (Stages 5–6)

Maintaining quality and controlling change are key facets of project management during the construction phase of any project. To what extent BIM has a role to play in these processes depends on the nature of the project and how it has been configured during the pre-contract phase in relation to information management. With a commission developed to comply with the PAS 1192:2013 standard, linkage between information required for construction and maintenance is explicit and wrapped around the definition and management of Project Information Models (pre-handover) and Asset Information Models (post-handover). For smaller projects, these may represent more theoretical than practical paradigms, depending on particular client interests and resources to manage their buildings from within digital environments.

According to Bob Garrett, the last few years have seen the rise of a new term, 'Field BIM', which is used to describe the use of BIM at the final stages of construction, particularly when commissioning and snagging are being undertaken, as well as when 'as-built' information is being gathered. Field BIM trawls information from digital models for verification and enhancement on site:

> Someone from the construction team might take a portable device, such as a tablet computer, into a room and confirm that all the information relating to the design and construction of that room had been carried out correctly. They could then note any discrepancies – which would add to the snagging list – or note any changes.
>
> (Garrett 2013)

The thrust with Field BIM is to use whatever tools are available to assist with the preparation of commissioning and use data to assist in the production of post-handover record information for the client.

Stage 5: Construction

- co-ordinate the use of digital data for contract administration purposes;
- maintain currency of BIM models during construction phase;
- initiate the recording of 'as-built' data for client use post-handover;
- where applicable, agree to relative scope and content of Asset Information Models;
- embed implementation of soft landings process into construction phasing.

Stage 6: Handover and Close Out

- agree implementation and auctioning of soft landings process (BSRIA Stages 3–5);
- issue record and maintenance data to client in agreed format (conventional O&M manual, online database resource, asset information BIM model).

Information exchange at Stages 5–6 (data drops 5–6) includes:

Drawings, schedules and agreed access to discipline-specific and federated BIM models where use is embedded into the contract conditions. Pre- and post-handover strategy documents for soft landings. Contractor-prepared record information as defined in the BEP. Delivery of Asset Information Models for client post-occupancy use (as applicable). Sustainability checkpoints 5+6 (carbon and soft landings strategies).

Note on Stages 5–6

The focus with Stage 5 is with execution of the building contract between site start and handover to the client team. This is the critical phase for delivering on the key drivers: completion on time, within budget and to the required specification.

Post-contract change management will be a key task for the client team in addressing variations to tender documentation arising either from predictable or unforeseen circumstances. If BIM models are referenced during the post-contract stage, the information manager's role continues in maintaining veracity and currency of data. Requests For Information (RFIs) need to be minimised and managed as they arise.

With a BIM project, there are a number of variables which could kick in at this point. For example, post-contract deliverables from BIM modelling environments could range from 2D orthographic drawing output in PDF format only to discipline-specific and/ or federated digital models for contractor/sub contractor use during the construction process; particularly if a project includes significant prefabrication elements where BIM models could be linked to manufacturing processes. For example, it is possible to link some MEP BIM software directly to the manufacture of ventilation ductwork.

If a soft landings process has been embedded in pre- and post-contract delivery, the BSRIA 5-stage protocol provides guidance on appropriate actions in the period prior to completion and in the immediate post-handover period (see Section 17).

15.5 Roles and relationships within design and construction teams

Unlike many sections of the manufacturing industry, built environment supply chain teams tend to be short-lived and transient. They may exist only through the time frame defined by a commissioned project's inception, development and delivery. Even for a large and complex building, that window of mutual engagement might be only 3–5 years maximum. For example, the technologically challenging 180m high 30 St Mary Axe skyscraper constructed by Skanska UK in the City of London commenced on site in 2001, was complete by 2003 and in occupation by 2004. By comparison, the original Boeing 777-200 passenger aircraft entered commercial service in 1995 and by 2014 was still in production; almost a 20-year window of continuity for improving and enhancing organisational techniques, including (perhaps most importantly) supply chain communications between service providers and their clients.

Central government's 2016 BIM initiative was primarily targeted towards the public sector where client teams are likely to include advisers with expert knowledge of built environment procedures and practices. Among the UK's constituent countries with devolved powers, Wales has already been active in addressing the client perspective for local authorities through the Consortium of Local Authorities in Wales (CLAW) BIM Toolkit.

The primary function of the tool was to provide Welsh local government commissioning bodies with a comprehensive suite of documents and tools to explain and facilitate deployment of a BIM-enabled approach in line with current UK standards and protocols. The Toolkit was also intended to support officers and members in articulating the BIM business case and efficient deployment and procurement practice to BIM-enabled projects.

Similarly in England, Kent County Council has been proactive through promoting the client role in 'pulling' supply chain organisations towards Level 2 BIM capability. In terms of local councils or other public agencies with the resources to assemble expert client groupings, a team might typically comprise representatives from the client organisation with the authority to make key decisions on project development, backed up by technical advisers with knowledge of BIM practice/procedures and building operations specialists with knowledge of specific information requirements for servicing post-occupancy needs.

Peter Trebilcock argued in 2014 that many clients were missing out on the benefits of BIM and cited a list of exemplars where engagement with BIM had demonstrably enhanced business performance:

- validation – the ability to interrogate design models in order to validate engineering decisions (Network Rail and HS2);
- show cost implications of design changes (power transmission);
- link directly with their current FM system (higher education, utilities);
- help eliminate interface and co-ordination challenges between packages/different teams (Crossrail);
- demonstrate safe methods of working, logistics planning and movement (airports);
- integrate data into their Asset Database (Highways Agency);
- provide linked schedules (Area, FFE, etc.) direct to models to ensure net lettable area visible at all times (commercial);
- provide detailed design information to aid Safety Case work (nuclear).

While not doubting the validity of the case made, the exemplars offered were all large organisations which would invariably embody client teams. For commissioning clients with more modest property portfolios as the likely employers of SMEs and micro organisations, the will and resources to feed the 'front end' of BIM projects (specifically EIR documentation) is less clear. Does the initiative come from the clients or the lead designers on that point? Or is an EIR's content likely to be teased out through dialogue between commissioning clients and their consultants?

At the time of writing, there was not a great deal of evidence that the BIM4 Private Sector Clients work stream set up by the Government BIM Task Group was active. From a cursory review of the literature early in 2015, it would seem there was still a great deal of work to be done on harmonising client aspirations with the use of digital technologies as enabling tools; certainly in terms of formalising engagement with the PAS 1192-2 protocol.

In the 2012 BIM Overlay to the RIBA Outline Plan of Work, editor Dale Sinclair identified that client representatives and their advisers would prepare a delivery plan at the commencement of a BIM project. This would be the primary instrument for setting out how project information was to be prepared (and by whom, when) and what protocols/procedures were to be used. The delivery plan would identify roles/responsibilities within the client team, and standard methods/procedures (including change control protocols). These could feed into the EIR once a team of consultants had been appointed to service a particular commission.

Organised and effective communication within built environment supply chain teams is a key ingredient for the successful project delivery. That imperative applies to every project, independent of the typology, scale or complexity of building. Mutual trust and respect among team members is a fundamentally important premise underpinning the interactions which feed into project development at every stage; from inception, design/construction through the occupancy phase of a building's whole life journey. Sirkka Jarvenpaa and Dorothy Leidner (2006) analysed and articulated the circumstances in which many organisations had formed virtual project teams which interacted primarily via digital networks. The nature of these 'new' socio-organisational structures identified a number of challenges which were open to question:

- Can trust exist in virtual teams where the team members do not share any past, nor have any expectation of future, interaction?
- How might trust be developed in such teams?
- What communication behaviours might facilitate the development of trust, particularly when team members had no history of working together?

The literature has variously cited the Integrated Project Delivery (IPD) methodology in relation to collaborative teamworking and BIM. According to the American Institute of Architects' (AIA) guidance for applying the IPD philosophy, the project team is formed as close as possible in time to the project's inception. If using a traditional procurement route for a BIM project, early interactions between design-construct team members may not always be possible because the process dictates linear rather than concurrent workflows. Contractor and sub contractor/specialist supplier knowledge which could usefully input to the setting up phase of working in digital environments may not be feasible in a competitive tendering situation.

The AIA argued that sometimes the project team will establish itself based on pre-existing levels of trust, comfort and familiarity developed through past working relationships. In other situations, the client may assemble the project team without any regard to previous relationships among the team members. When engaging with and embedding 'new' working practices like BIM, previous positive experience across team members could be perceived as potentially enhancing interactions. Conversely, the situation where a project team which mixed BIM-experienced organisations with firms which were BIM novices could be inhibiting or even problematic in terms of overall team performance.

It has been argued that construction project teams include two categories of team member: the primary participants and key supporting players. The primary participants were defined by the AIA as 'holding roles' which had substantial involvement and responsibilities throughout the project. For example, with a traditional procurement route, the primary participants would typically be the client, design disciplines and contractor. The research suggested that unlike the relationship in a traditional project, the primary participants in IPD may be defined more broadly and are bound together by either a contractual relationship (as would be the case if the CIC BIM Protocol was applied to a project), or by other less tangible drivers, such as client relationships, ethical issues and the like.

Making reference to the earlier exemplar, key supporting participants (for example, an energy consultant) on an integrated project fulfil vital roles on the project, but perform more discrete functions than primary participants. In a 'traditional' project, the key supporting participants include the primary design consultants and subcontractors. In IPD, the key supporting participants enter into contracts directly either with one of the primary participants or with the commissioning client. In either event, all parties agree to be bound by the collaborative methods and processes governing the relationship among the primary participants. In IPD, the difference between the primary players and key supporting participants was noted as being a fluid distinction which may vary from project to project.

The 2012 BIM Overlay to the RIBA Outline Plan of Work noted that establishing methods for structuring model information within design and construction teams would be critical to the success of the next generation of BIM projects. However, design leadership would be of greater importance. In order to work collaboratively, new approaches to teamwork would only be successful if they were aligned with redefined discipline roles and responsibilities.

Without this clarity, it would not be possible to adequately define shared working methodologies which were free from ambiguities when engaging with BIM environments. As has been mentioned elsewhere in the book, UK built environment discipline roles/identities are well established and there is no significant evidence of these being re-mapped in the short to medium term to align with new ways of digital working. There are a number of BIM postgraduate programmes available in the UK, but these tend to build from discipline identities already established through undergraduate education and training.

As Dale Sinclair also noted, the *CIC Scope of Services Handbook* does provide guidance on definition and use of the integrated and detailed scopes of services for use by members of the project team undertaking building projects (but is not BIM specific). The terms, design manager, project manager, information manager, model manager et al. tend to be used quite loosely and sometimes without qualification with regard to discipline roles in collaborative projects.

Even small commissions will demand overall design leadership as a co-ordinating role. On larger projects using a traditional procurement route (such as the exemplar discussed above), each contributing discipline will have a lead designer. Projects which have early contractor/subcontractor involvement may be directed and co-ordinated by a project and/or design manager.

The term 'information manager' is frequently used in relation to BIM. The CIC BIM Protocol incorporates the role into the document's rubric and distinguishes between information and design management. However, according to the RIBA Outline Plan of Work BIM

Overlay, there was lack of clarity regarding this role, but if the lead designer role remains with BIM projects, it is important that the information manager's role does not conflict with design responsibilities, either at discipline level or in an overall co-ordinating role.

One interpretation was that the information manager takes overall responsibility for managing the inputs of each designer into the project model. There seems no particular discipline associated with fulfilling the information manager role at this point in time. The tradition with UK built environment undergraduate programmes is to provide the industry with generalists across the various established discipline remits.

With a Level 2 BIM project developed to harmonise with the PAS 1192 suite, it would seem reasonable to expect that the designated 'information manager' would co-ordinate all aspects of the setting up of protocols and management of digital data across the design-construct team, including managing the CDE. The information manager's affiliation within the project team could depend on the commission's size and procurement type.

As was discussed elsewhere, for a small project using a traditional procurement route, the information and design management roles could be carried out by the same person. For larger projects, for example procured using a design-and-build methodology, the information manager could be embedded within the contractor's organisational structure and would be likely to have an overarching role which extended across the project supply chain, from design team members through the main contractor to specialist subcontractors and suppliers.

Interestingly, one of the UK's leading BIM recruitment specialists does not use the terminology 'information manager' in describing job roles associated with the manipulation and co-ordination of digital data. They use a three-tier designation, which ranges from 'Modeller' (technician/graduate role) through 'BIM Co-ordinator' (technical/professional project specific) to 'BIM Manager' (a role which suggests a more strategic level within an organisation, with responsibilities for strategy, research, process/workflows, standards, protocols and audit).

Over the last few years, the BIM Manager designation seems to have been on the ascendancy and has become a feature within the organisational hierarchies of many built environment companies (including SMEs). These range from design firms, surveyors through contractors/subcontractors to specialist manufacturers. Also, in relation to a role tagged as the 'Construction process engineer', the construction industry demands new skills to do BIM and these fresh skillsets need to cross traditional discipline boundaries. Construction Process Engineers are required to be 'versed in 3D modelling, design, engineering, construction technology, quantity surveying, be knowledgeable about databases, and have the technical skills to resolve difficulties at the interfaces between processes' (Barker 2011).

15.6 Strategies for cost control and client interactions

Dick Barker, Head of Model Based Measurement and Costing at Laing O'Rourke, argued that 'most people in the industry will now have their own definition of BIM, and the majority will have a 3D model and data in there somewhere' (Barker 2011). According to Louise Sabol, BIM offers a potentially 'transformational technology' through its capability to provide a shared digital resource for all participants in a building's life cycle management, from preliminary design through facilities management (Sabol 2008). As a visual database of building components, BIM can provide accurate and automated quantification, and assist in significantly reducing variability in cost estimates.

Barker also argued that, in considering BIM from a measurement and costing perspective, there are three important cornerstones underpinning the process:

- 3D modelling to generate the geometry (and attached metadata) of what has been or might be designed;
- defining the scope of work in bills of quantities and dynamically populating the bills with quantities from the model/s;
- specification and auto-annotating the 3D and 2D drawings from the model, as well as dynamically linking the bill descriptions to the specification.

At NRM1 (order of cost estimating and cost planning) stage, order of cost and cost target exercises for budgeting purposes can be undertaken from a combination of site information designers' (all disciplines) conceptual sketches and historical cost reference data. It is a time-honoured tradition in the UK that sufficient information is provided at early stages of project development to support the decision-making process only. As that situation does not change with BIM, interaction with schematic digital models at early stages of project evolution is not a significant issue with order of cost planning. However, in following a key principle that digital information should only be used once if possible, where schematic models are available and embody realistic and validated geometries, they can be 'mined' to determine floor areas, building footprints and indicative storey heights.

To apply NRM2 (detailed measurement for building works) will invariably involve interaction by the project quantity surveying team with BIM models. As was discussed, data can be extracted 'automatically' using specialist software such as CostX. For smaller projects, areas, volumes and quantities can be extracted manually from discipline models using interactive viewers. Typically, these are free software downloads and generally offer adequate to good functionality. As a partner from a national surveying practice commented, despite the time-saving advantages of being able to number-crunch quantities from BIM models quickly and efficiently (as has always been the case), the issue of accuracy of information is often still in the frame. Also, in the situation where design disciplines produce 'own models' which are combined in a federated model environment, levels of detail or development represented at specifc work stages can vary by discipline.

At NRM3 level (order of cost estimating and cost planning for building maintenance works), cost plans can be integrated with maintenance and life cycle replacement works. This raises the possibility of cost data being embedded in BIM models and carried forward for whole life use. NRM3 also gives direction on how to quantify other items associated with building maintenance works. As has been noted, these are not necessarily reflected in the measurable maintenance work items, for example 'maintenance contractors' management and administration charges, overheads and profit, consultants' fees, employer definable other maintenance-related costs; and risks allowances in connection with maintenance works' (Green 2014).

Unlike capital expenditure, which is relatively short-term, maintenance works are required to be carried out from the day a building is occupied until the end of its service life. Accordingly, while the costs of a capital building works project are usually incurred by the building owner/developer over a relatively short term, as MacLeamy noted (Section 18) costs in connection with maintenance works are incurred throughout a building's whole life. In that context, NRM3 also provides guidance on the measurement of whole life factors which can influence cost planning and can be linked with digital databases.

Industry feedback suggests that data exchange between disciplines in collaborative environments can be problematic. For example, three of the most popular BIM authoring software packages all use different native file sources. Also, some proprietary software updates are not backward-compatible, meaning that even native files authored by the same BIM application (but different releases) may not be readable. Exchanging data between software platforms may also be challenging if discipline design teams are using different applications which rely on the 'open' IFC standard to be able to read and interpret data. These difficulties can be exacerbated by output inconsistencies across file exchanges and by the need to apply 'fixes' on the fly to stabilise performance with software applications.

As Sabol noted, in fact, a two-way exchange ('round-tripping') of BIM geometry between formats (e.g. native file type 'X' out to IFC and back to native file type 'Y') is, as of yet, problematic. A process which could exchange and update geometry and integrate data-driven changes is beyond current BIM capabilities (certainly in terms of software generally available to design and construction disciplines).

The capability for BIM-generated cost-estimating applications to update and change the original building model remains undeveloped, according to Louise Sabol. Costing exercises often result in the need to substitute components of design at different stages of project evolution. Applications have not yet developed the facilities to export these decisions back to BIM and automatically update the building model to reflect the changes. Current costing software packages do have the ability to flag items which have either been changed or have been added to the original model, so that estimators can modify and update the object base used for cost planning purposes.

Generally, possibilities for client interactions with BIM models are quite limited at present, particularly the facility to 'read' the effect of design modifications on cost planning in real time. The most significant challenge for hands-on 'dialogue' is the need to be able to engage with and manipulate data-rich digital models (the alternative is to use a model viewer). In that context, cost reporting would be among BIM outputs delivered in more traditional formats, via spreadsheets or online PDF files supplied at agreed information exchange points (identified in a project BIM execution plan). As augmented reality applications gradually come to market and become more widely available, new opportunities for deeper levels of client engagement with digital processes (to facilitate decision-making) could start to impact on work practices.

Bibliography

Anon (2007a) *Integrated Project Delivery: A Guide*, American Institute of Architects, California Council. Version 1, [online]. www.aia.org/groups/aia/documents/pdf/aiab083423.pdf [accessed 24 March 2015].

Anon (2007b) *The CIC Scope of Services Handbook*, 1st edn, Construction Industry Council, [online]. http://old.cic.org.uk/cicservices/CIC_ServicesHandbookOct07.pdf [accessed 26 March 2015].

Anon (2011) CIC Consultants' Contract Conditions, 2nd edn, Construction Industry Council, [online]. http://cic.org.uk/publications/?cat=contracts [accessed 26 March 2015].

Anon (2013) *RIBA Plan of Work 2013*, RIBA, [online]. www.ribaplanofwork.com/Default.aspx [accessed 24 March 2015].

Anon (2014a) 'New Rules of Measurement', *Designing Buildings Wiki*, [online]. www.designingbuildings. co.uk/wiki/New_Rules_of_Measurement [accessed 24 March 2015].

Anon (2014b) 'RICS NRM: New Rules of Measurement', RICS, 7 November, [online]. www.rics.org/ uk/knowledge/professional-guidance/guidance-notes/new-rules-of-measurement-order-of-cost-e stimating-and-elemental-cost-planning/ [accessed 24 March 2015].

Anon (2015) *Managing Health and Safety in Construction*, Construction (Design and Management) Regulations 2015, Health and Safety Executive, [online]. www.hse.gov.uk/pubns/priced/l153.pdf [accessed 24 March 2015].

Anon (n.d.) *30 St Mary Axe*, Skanska UK, [online]. www.skanska.co.uk/upload/Sevices/Design/30%20 St%20Mary%20Axe.pdf [accessed 24 March 2015].

Barker, D. (2011) 'BIM – Measurement and Costing', *Building Information Modelling*, NBS, February, [online]. www.thenbs.com/topics/bim/articles/bimMeasurementAndCosting.asp [accessed 24 March 2015].

Earl, S. (2012) 'The RICS' New Rules of Measurement', *Building*, 23 March, [online]. www.building. co.uk/the-rics-new-rules-of-measurement/5033890.article [accessed 24 March 2015].

Garrett, B. (2013) 'BIM for Handover', *CAD User*, May, [online]. www.btc.co.uk/Articles/index.php? mag=Cloud&page=compDetails&link=2496&cat=CAD [accessed 26 March 2015].

Green, A. (2014) *NRM 3: Order of Cost Estimating and Cost Planning for Building Maintenance Works*, RICS, [online]. www.rics.org/uk/shop/NRM-3-Order-of-cost-estimating-and-cost-planning-for-building-maintenance-works-19866.aspx [accessed 26 March 2015].

Hamil, S. (2015) 'The Toolkit for BIM – Completing the Jigsaw', Adjacent Digital Politics Limited, 24 February, [online]. https://www.adjacentgovernment.co.uk/housing-building-construction-planning-news/toolkit-bim-completing-jigsaw/ [accessed 24 March 2015].

Jackson, R. (2015) 'The NBS BIM Toolkit Public BETA Feedback', *BIM Blog*, Bond Bryan Architects, [online]. http://bimblog.bondbryan.com/the-nbs-bim-toolkit-public-beta-feedback/#more-2620 [accessed 29 April 2015].

Jarvenpaa, S. and Leidner, D. (2006) 'Communication and Trust in Global Virtual Teams', *Journal of Computer-Mediated Communication*, 3(4), 1998, [online]. http://onlinelibrary.wiley.com/doi/10.1111/j.1083–6101.1998.tb00080.x/full [accessed 26 March 2015].

Ravenscroft, T. (2015) 'NBS Toolkit: Is it BIM's Game Changer', *BIM+*, CIOB, 22 March, [online]. www.bimplus.co.uk/news/nbs-toolkit-preview-it-bims-game-changer/ [accessed 26 March 2015].

Sabol, L. (2008) *Challenges in Cost Estimating with Building Information Modeling*, Design + Construction Strategies LLC, [online]. www.dcstrategies.net/files/2_sabol_cost_estimating.pdf [accessed 26 March 2015].

Sinclair, D., ed. (2012) *BIM Overlay to the RIBA Plan of Work*, RIBA Publishing, [online]. www.architecture.com/files/ribaprofessionalservices/practice/general/bimoverlaytotheribaoutlineplanofwork2007.pdf [accessed 24 March 2015].

Trebilcock, P. (2014) 'Clients Are Missing Out on Half the Benefits of BIM', *Building*, 25 March, [online]. www.building.co.uk/clients-are-missing-out-on-half-the-benefits-of-bim/5067367.article [accessed 26 March 2015].

16 Pathways from design towards construction and use, inputs and outputs

16.1 Shades of BIM, lonely BIM, collaborative BIM *et al.*

One of the practical difficulties of putting the BIM acronym on a level footing of understanding for practitioners, educationalists and researchers is finding some generally accepted universality of meaning. Did the term originate with a 1992 paper published in the *Automation in Construction* journal (van Nederveen and Tolman), as has been claimed; does the third letter of the acronym refer to 'Modelling', 'Management' or both? The huge contribution that Charles Eastman of Georgia Institute of Technology has made to BIM research and associated academic development over many years began in the mid-1970s. By the time Autodesk's white paper on 'Building Information Modelling' was published in 2002, ubiquitous BIM authoring packages like ArchiCAD (1987) and Revit (1997) had been around for at least five years or more. In 2015, from a cursory trawl of information published on company web portals, many SMEs and micros are now claiming to do BIM, but the detail on collaborative practice (particularly application of BS1192:2007 and the PAS suite) is more difficult to flesh out.

The literature suggests that the 'BIM' acronym has been subject to various nuances of origin, scope and interpretation for some time. Some abhor the acronym and say the sharp focus should be on how design/construct/maintain teams manage digital information in virtual environments. People/process, inputs, outputs and cross-disciplinary dialogue (BIM or otherwise) remain central to the process of making buildings. Since the UK Government's 2011 Construction Strategy was published and the 2016 BIM threshold was announced, BIM has seldom been out of the built environment industry's news threads as a topic for conferences, seminars, webinars, blogs, CPD events, and the like. Meanwhile, the 2015 NBS industry survey suggested that for UK industry BIM uptake had plateaued or might even be on the wane.

Over the last few years it seems like nearly every possible aspect of BIM has been unpicked and chewed over in terms of trying to move forward agendas for change. For construction, the primary quest is for more efficient information management in pitching towards achieving productivity, efficiency and other gains. However, finding demonstrable unanimity of purpose across the built environment sector has been extremely difficult, because of the way that the industry is structured and operates. Nearly every built environment professional body has developed its own take on BIM, but evidence of meaningful cross-disciplinary collaboration between the institutes is rather thin. Perhaps it is significant that when architect Stephen Hodder spoke at the 2014 CIBSE conference, it was reported he was the first RIBA president to do so.

The BIM acronym is used in various contexts to describe different sets of circumstances. In some ways, that situation is not particularly helpful, particularly for SMEs and micro organisations which might be considering taking first steps towards engaging

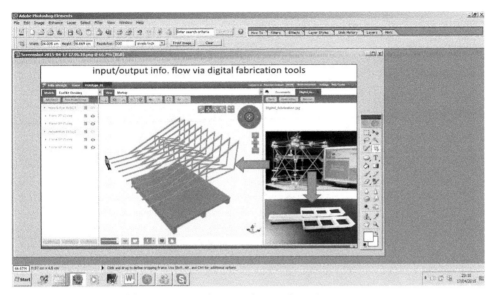

Figure 16.1 Using digital data to connect design with manufacturing processes

with digital environments to support the development of construction projects. As with cross-disciplinary collaboration, there is not a great deal of evidence from the literature that the BIM Task Group's flagship standards (PAS 1192 suite) are being applied by SMEs and micro organisations en masse in ways which might signal a significant force for industry change. Compare that situation with statutory instruments like the revised Construction (Design and Management) Regulations, which came into force in April 2015 with immediate effect on construction in its entirety.

From other perspectives framed by contemporary settings, it really doesn't matter that BIM lacks definition and has fluidity with interpretation. Judging by the output from prolific BIM bloggers like Stephen Hamil of the NBS and Rob Jackson of Bond Bryan Architects, there is general and tacit acceptance that doing BIM involves using data-rich authoring software and varying levels of collaboration between participants engaged in the temporal processes of making, using and managing buildings. It seems that BIM has not yet facilitated great gains made in connecting building design with manufacture (Figure 16.1). In addressing the challenges raised by Sir John Egan in 1998, analogies with manufacturing industry paradigms seem as far removed from mainstream design and construction practice as 20 years ago:

> Last week I attended the Associated General Contractors of America (AGC) BIMForum – a focus group for contractors implementing Building Information Modeling. The group meets quarterly and features John Tocci, Sr. of Tocci Construction of Massachusetts presiding over the intense 2-day agenda of presentations, case studies, breakout workgroups and 'rapid-fire technology demonstrations.' A conference of this nature differs from the likes of Autodesk University or the Bentley Empowered Conference in that one won't find 'how to' classes, rather industry professionals – including architects, engineers, subcontractors and fabricators – sharing their experiences and crafting solutions to today's and tomorrow's BIM challenges.

> (Van 2008)

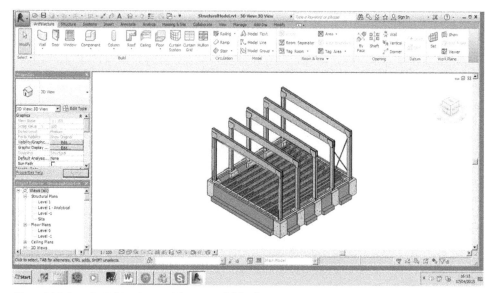

Figure 16.2 Lonely BIM, sole practitioner model for house extension in Autodesk Revit. Functionality
restricted to capturing orthographic model views for planning/construction purposes

James Van's blog from the 2008 AGC BIMForum, held at a venue outside Washington DC,
is interesting on a couple of levels. First, it emphasised the value of shared experiences with
BIM (as a learning process) across construction discipline 'boundaries'. Van also reported that
John Tocci introduced a new nomenclature of 'lonely BIM' versus 'social BIM' and noted
these shades of BIM drew from Finith Jernigan's 2007 book *BIG BIM, little bim*. In his 2013
blog post 'There are Four BIM Flavors', Jared Banks described a hierarchy of BIM scenarios
(hypothetical and otherwise) which develop from lonely little BIM (Figure 16.2) through
social little BIM towards lonely BIG BIM to social BIG BIM as the ultimate expression of
BIM community interactions.

As with any culture-challenging shift like BIM, people quite naturally strive to flush out and pin
down tangible and definable concepts. That exercise may well assist with gaining knowledge and
understanding of new things and work practices. However, it does seem odd that with data-rich
architectural authoring packages having been around for at least 15 years (by 2008, for example,
ArchiCAD had reached Version 12), it took people so long to latch on to the idea that it was the
data embedded in these packages which needed to be unlocked, filtered and moved around design-
construct teams. Plus, increased emphasis on whole life aspects of building performance up to the
point where a building might be decommissioned and safely dismantled/recycled stretched the
conceptual and practical time frames over which data might usefully be deployed.

There is no huge gain for a small organisation in dwelling on and trying to unpick the
perceived shades of BIM in great detail. However, it might be useful for SMEs and micros to
consider how these various scenarios could help clarify what doing BIM means at a practical
level, first, as a self-appraisal exercise within an organisation in relation to current and ongoing
workloads; and thereafter, in reaching out and reviewing relationships with partner organisations with regard to BIM implementation. Developing these exercises at a practical level
would necessarily involve thinking about business size and structure, project portfolio, client
base and typical fit of the organisation within design-construct teams.

A common interpretation of the lonely BIM moniker is when used to enhance operational efficiencies production gains within a single organisation. A case in point could be a structural engineering firm or architectural practice using BIM authoring software to generate a data repository for commonly used construction objects, specification clauses and the like, beyond the immediate needs of servicing an ongoing project.

At the other end of Jared Banks' spectrum of BIM flavours, 'social BIG BIM' implies the process of generating data for sharing across a range of disciplines during the building design, construction and operational phases. Applying the PAS 1192-2 protocol to that paradigm suggests that in project-specific instances, the primary facilitator for data sharing would be a Common Data Environment (CDE), supplemented by appropriate design/project management software tools to assist design/construction team interactions as data accumulates, flows and changes through iterations across project life cycles. As has been suggested earlier, perhaps the acts, manifestations and consequences of digital modelling (and models) are subservient to bigger picture and temporal processes for data management, particularly in considering the range of organisations making up a supply chain network and the ability of the smaller players to fit the bill, engage and deliver required outcomes.

For example, in considering the supply chain structure for a £25 million school project developed on a Level 2 BIM template, a suspended ceiling specialist bidding for a £20k sub-contract might have no knowledge or skillbase with BIM authoring software, but would be expected to buy-in to the data sharing/exchange process developed via the CDE. The necessary skillsets/protocols would be written into the project BIM Execution Plan and tested, for example, during a prequalification process using the PAS 91:2013 template.

John Adams, BIM project manager at 4Projects, tackled that issue of supply chain inclusion by making reference to subcontractor interactions with BIM environments. In addition, Adams addressed perceived challenges with information 'diffusion', specifically data flows across supply chains and accessibility on a 'need-to-know' basis. Also, in situations where BIM models exist within CDEs, how participants engage with information exchanges across design-construct teams using email, Dropbox, Google Drive and similar cloud-hosted alternatives to FTP servers.

To address these challenges, 4Projects initiated the 'Tier2Tier' research and development exercise, which was undertaken by a consortium including Vinci Construction, M&E contractor NG Bailey, the Specialist Engineering Contractors' Group and the University of Northumbria. The central idea was that subcontractors working on major BIM projects could more efficiently interact with the 'main BIM model' via their own satellite CDEs. As Terry Gough, Kent County Council BIM Champion, pointed out, the ethos underpinning that premise may challenge the central idea of a CDE as a centrally held repository for digital data accessible to the project team. But then as Mervyn Richards noted in his guide to BS 2007, a CDE is a 'procedure', not a thing. As with other aspects of BIM practice, as ideas and standards evolve, it may become problematic if concepts which have not been extensively road-tested over time are regarded as fixed reference points.

The 4Projects cloud-based service was intended to allow key subcontractors to operate a satellite CDE, which would only host data from the main model relevant to their package, for example M&E. The subcontractors would then be able to invite suppliers and manufacturers to access the 'slimmed-down' design files and submit their own designs, estimates and COBie data, which could then be uploaded into the 'central' project CDE. Adams argued that in terms of extending supply chain engagement, 'It's a more collaborative approach to building up the data' (Knutt 2015b). Each participating organisation would have its own 'satellite', where organisation-specific BIM information could be checked against the BIM Execution

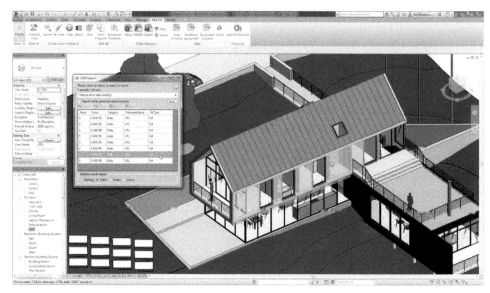

Figure 16.3 Loughborough University-developed 3DIR plug-in for Autodesk Revit. Allows critical data to be extracted from model using keywords as prompts

Plan and file-naming conventions. The design/supply/construct team would all work to the same protocols, but specialist subcontractors would be able to operate from within their own project-specific digital environments.

Another aspect of collaborative/social BIM which had started to gather momentum at the time of writing was the issue of information overload through engagement with data-rich BIM models. For example, in April 2015 construction company Mace announced it was prototyping a user-friendly 'SlimBIM2go' platform for clients who want digital FM solutions without the need to get to grips with data-heavy as-built models BIM and COBie. The SlimBIM2go approach takes an 'as-built model' from the construction team, then filters data to produce a tablet app which allows users to navigate the building digitally without the need to engage with data-rich BIM authoring software. Users can manipulate constituent layers of the finished building, with hyperlinks to more detailed information sources.

Also in 2015, researchers at Loughborough University developed a plug-in (3DIR) for Autodesk Revit (Figure 16.3) which allowed the design-construct team member to search a 3D building model for information by using specific keywords or specifying a 2D area, 3D volume or set of 3D components. Search results were displayed either as a list or could be superimposed onto the 3D model itself. Relationships between 3D objects were also used to rank search results. 3DIR worked on the principle that the 3D visualisation or 3D data can be exploited when running search queries, delivering more tailored, digestible and engaging results than those produced, for example, using a web search engine.

To paraphrase thoughts from Dr Peter Demian from the School of Civil and Building Engineering at Loughborough, BIM is very important for construction, and more information is being crammed into 3D BIM models, making the hunt for relevant data increasingly challenging:

> Our research shows that when faced with a situation of information overload such as this, the 3DIR app is a lot more effective in engaging the brain than text based systems such

as web search engines. This is largely due to the fact that the app is picture-led and links together graphical and non-graphical information in an easy-to-read format.

(Anon 2015a)

Sean Benson argued that learning how to use the parametric and structured data stored in BIM models appropriately may be key both to shifting from 'lonely' towards the 'social' end of the spectrum of commonly described BIM paradigms and also to raising client awareness of how data embedded in BIM authoring packages can be unlocked, shared and usefully deployed post-occupation for building management. Again, the emphasis was on identifying the scope of digital data required to service project needs in the transient journey from design through construct to operational phases. Ultimately, that data could also be useful to service end-of-life requirements, such as deconstruction and recycling of components, assemblies and whole buildings.

Benson cited Wisconsin's Department of Administration (DOA) as offering an early adopter example of client-driven imperatives impacting on the procurement process and BIM methodologies. DOA project manager Bill Napier charted Wisconsin's BIM journey as beginning in 2006 and being triggered by challenges posed by the enablement of energy management strategies. With a building estate comprising more than 6,200 facilities (combined replacement value estimated at £6.16 billion), it quickly became clear that BIM would be a necessary enabler to meet requirements of the executive's £0.78 billion annual portfolio of construction projects involving more than 500 design-construct firms serving 16 state agencies and a 13-campus university system.

In 2009, Wisconsin's Division of Facilities Development published a handbook describing in detail requirements for the selection of consultants as service providers for the State's property portfolio. Procurement methodologies for all new buildings above a threshold value of £1.6 million were required to incorporate BIM. Wisconsin's template set out standards for the capture and sharing of digital data and must be embedded in any new construction projects above the fiscal threshold. The detail of the State's standards reflected the perspectives of three key players in procurement processes: building owners/operators, designers and constructors. By any measure, these protocols defined BIM scenarios which included the full gamut of supply chain organisations within their reach. Even the smallest construction organisation was required to be BIM compliant to be included in the frame of reference.

16.2 Appraising and selecting appropriate paradigms for design management

A great deal has been written about BIM process and standards in relation to information management. Less so about design management, the broad and often generic term which is applied across industrial sectors. Within construction, the design management function is well established. In terms of the UK scene, John Eynon attributed that situation jointly to establishment of the design and build procurement methodology and rise of specialist subcontractors as key players in contemporary construction genres.

As a strategic function within organisations, design management typically aspires to link design, innovation, technology and management in pursuit of specific goals. For example, when the BBC relocated 1,600 media jobs from London to MediaCityUK at Salford Quays, the original business case identified a need for 80,000 square metres of floor space. However, by evaluating the BBC's business process and work styles, deriving and applying best practice work settings, occupation densities and use of studio facilities, consultant Capita's project team were instrumental in being able to reduce the accommodation by over 50 per cent resulting in significant capital and operational cost savings.

At project level, design management may be perceived as part black art and part science (an amalgam between subjective and objective decision-making) acting as a catalyst to enhance collaboration and the attainment of operational synergies within design and construct teams. Design managers typically come from a design background and combine knowledge of design and construction process with exceptional people and co-ordination skills.

Many large construction companies, like Skanska, Brookfield Multiplex, Kier, Lend Lease and others, employ design managers to ensure that an effective and productive liaison is established between design, construction and client teams. The design manager needs to be able to bridge between two cultures (the divergent/discursive characteristics of design and construction's convergent pragmatism) in driving projects forward towards attainment of client goals. Managing change, including embedding requests for information into change processes, can flag up significant challenges for design managers, particularly once a project (or design packages making up part of a project) has been fully documented and issued for construction purposes.

Andrew Barraclough, Group Design Director at Wates, one of the UK's largest speculative housebuilders, commented:

> some would argue that contractors simply offer input on buildability and sequencing which has value but doesn't really contribute to the design. But contractors need to have experienced staff from architectural backgrounds if they want to make a difference to this complex process and ensure the vision for the build is translated throughout the process.
>
> (Barraclough 2015)

Typically, on large projects, the design manager's principal responsibility is to co-ordinate the multidisciplinary team to ensure that design intent is realisable within time and cost constraints, and to ensure the integrity of the design concept is maintained through design development and construction phases of project development by troubleshooting, resolving and managing the issues which invariably will arise with any construction project. Whatever the scale of project, design managers need to be able to:

- develop a deep understanding of the client's design brief;
- establish and manage an effective working environment for the design-construct team;
- develop and manage the design development programme;
- liaise between client and design-construct teams in managing design decision-making and change issues likely to impact on time/cost/quality;
- ensure that discipline-specific design information is validated, co-ordinated and integrated into project development;
- embed collaborative decision-making into the management of project information/ document control systems;
- identify and manage the design-related risks as projects evolve;
- ensure buildability at all stages of project evolution/development;
- act as a point of contact in managing requests for information and design variations; and
- manage the compilation of design-related information required to service clients' post-handover requirements.

The generic function of design management is applicable at any scale of project realisation, yet only major projects may justify the appointment of a design manager as a separate role. For SMEs and micro organisations, the development of small commissions will typically subsume design management functions into discipline-specific activities, with a nominated design-construct team member taking overall responsibility for design management and co-ordination. That role is often taken on board by the lead designer.

There is a culture prevalent in construction that, after a certain stage in design evolution, change becomes disruptive and likely to have a negative impact on time/cost/quality. For example, the clients for a single family house asking their builder for a window position to be altered two weeks before handover. But that viewpoint is balanced by a contemporary counter-culture which argues that the ability to respond quickly and efficiently to client-driven change scenarios can add value to customer perceptions of design-construct team performance. In an environment where client business needs were constantly evolving, Finnish furniture company Fira used cross-pollination between manufacturing (Big Room/Concurrent Engineering) and construction (BIM) paradigms to put that premise to the test.

Since 1996, construction-specific guidance for design management has been offered by BS7000–4 (updated in 2013 to include BIM standards). The December 2013 version cross-refers with BS1192, PAS 1192-2 and BIP 2207, the standard framework and guide to BS1192–2010. BS7000-1 2008 is a separate document in the BS7000 series and provides guidance specific to managing innovation – a useful reference document, particularly for BIM scenarios, where introducing innovation to process may heighten risk. For example, when teams are using unfamiliar web-based tools on a live project to track and feedback on design development. BS 8536-1 2015 may also be relevant in linking pre and post occupancy aspects of project development.

From a design management viewpoint, what are the differences between a commission documented along conventional lines and the same project undertaken using BIM methodologies? In addressing that question, there are three strands of activity to consider: core values, procedural issues and cultural aspects. Invariably, in practice these themes may not be mutually exclusive and will overlap.

The core values for design management apply universally, independent of the method/s used for documenting project development. These imperatives are underpinned by the need (whatever the scale of the project) to embed/apply robust and effective protocols for collaboration and management of process in the quest to achieve desired outcomes. From a client's perspective, there is no inherent difference between a single family house subject to a 25 per cent cost hike between detail design phase and occupation, and a £20 million commercial development where the outturn cost rises to £25 million. Neither project has delivered on client expectations.

In terms of procedures, if client imperatives demand implementation of the PAS 1192-2 protocol, for the SME or micro organisation providing the design management lead, delivering on that requirement will invariably involve an additional raft of project documentation. Is that a challenge or an opportunity? Probably both. At this point in time, there seems no other way for effective and reliable engagement with Level 2 BIM in the UK. Over time, the evolution of variations and/or addenda to PAS 1192-2 may assist in tailoring the standard into leaner and bespoke forms for small projects.

While taking the scope of the standards on board, interpreting and tailoring PAS 1192-2 into project-specific formats may be time hungry for SMEs and micros. Resourcing that overhead will need to be factored into fee bidding. Anecdotal evidence suggests that it may be more difficult for smaller organisations to absorb the BIM factor than larger firms. For example, in the case of a project which combines a traditional procurement route and separate consultant appointments with a single-stage tender process.

In developing project documentation and considering how digital data will be embodied into process, the design and information management roles will invariably merge on a small project. However, the consequences of change still need to be managed (for example, in the case of a single storey nursery school where the client has asked for the internal room arrangement to be altered during the detail design phase). What are the time and cost consequences of embodying change? Are they different if the project had been documented using 2D CAD compared with a federated BIM model scenario?

Beyond considerations bound up in process, cultural aspects (predominantly people issues) are embedded in application of the design management role to BIM. Engagement with unfamiliar tools and procedures may heighten risk as a result of interactions between design-construct team members. Typically these 'soft' people facets of design management and BIM will involve getting to grips with the challenges of collaborating in digital environments, collective use of online tools for design reviews, engaging with hybrid processes, for example toggling between 3D federated models for design development review combined with co-ordinating flat 2D drawings which may be used for contractual purposes.

Managing human aspects of addressing these challenges will fall within the design manager's remit as a co-ordinator of people and process on a BIM project. Unfortunately, with reference to the 20 per cent technology and 80 per cent people/process BIM paradigm regularly cited by commentators, PAS 1192-2 and its partner documents do not extend to providing guidance on the human/cultural factors embedded in the mix. That remains a perennial issue which the construction industry has yet to address effectively in moving forward with collaborative BIM scenarios. It is not enough just to define new roles (for example, information manager) in relation to BIM environment practice. These roles need to be capable of being comfortably assimilated into all organisational activities at all levels.

16.3 Classification systems, CPIC and Uniclass

From the authors' experience, two topics guaranteed to raise eyebrows among SME practitioners during discussions on BIM semantics and practicalities are classification systems and COBie. Yet exploring and applying both is fundamentally important to developing legible, understandable and consistent roadmaps for digital data in transit through interactions with BIM models. When BIM enters the frame of reference, it impacts on a wide spectrum of 'things', from individual construction components through spaces, assemblies, buildings and facilities to infrastructure. Associated temporal movement of digital information may occur within project teams and across disciplines and construction industries, both nationally and internationally. At project level, inputs and outputs are mapped and controlled by protocols like project BIM Execution Plans. Feedback from small organisations suggests that these paradigms may be viewed as theoretical templates and do not necessarily reflect the realities of workplace experiences, particularly with small projects.

Digital data exchanges weave a complex web of information sources and labels which need to be structured logically and consistently to avoid confusion, multiple interpretations and diversity of meanings. If that sounds like a somewhat idealistic perspective, that was not the intention, but does highlight challenges in harmonising between regional and international standards for built environment classification. Ironically, many small construction organisations may have little day-to-day engagement with classification. But for others, particularly component manufacturers and specialist suppliers (for example, of facade systems), appropriate labelling of their products to fit with contemporary BIM practice may already have become central to 'now' and forward-looking business development challenges.

Pre-BIM, the most widely used classification typologies were work sections for specification writing and elements used for cost planning/analysis. However, the ascendancy of freely available BIM objects (both generic and product specific) has upped the ante in terms of defining properties and characteristics of construction products generally. Many product manufacturers now offer digital representations of ubiquitous objects, such as doors, windows, insulation and roof tiles, in BIM formats with embedded metadata which can be used both for specification and post-occupancy maintenance purposes.

Early in the 1990s, International Construction Information Society (ICIS) members initiated a journey for regularising national construction classifications into a universally recognised international protocol. These discussions led to the formation of the ISO 12006-2 standard which Anders Ekholm noted had its roots in the Swedish SfB system. Since its formation, ISO 12006-2 has been applied in the development of building classification protocols like the Swedish BSAB 96 standard, the UK Uniclass (later superseded by Uniclass 2), North American Omniclass and Danish DBK-system.

At ISO meetings in Vancouver in 1999, a variety of organisations developing IT standards for the building industry agreed that some sort of standardised global terminology was necessary and that its structure needed to facilitate the reliable exchange of data between computers, irrespective of semantic language. Modern information systems for construction need to be able to handle digital data with different characteristics: geometric properties, technical specifications of objects, maintenance data, and the like. BIM-related software is diverse and ranges from design/management tools to systems for servicing specification, product data and cost information system needs. All these data streams and their relationships and interdependencies, need to be defined and structured in such a way that the stored information is robust, consistent and reliable within and between the different applications. Also, data needs to be able to move about in digital environments without loss of key characteristics. In the case of BIM objects or assemblies, for example, a common scenario would be the need to transfer files across different software platforms which use different native file types.

The international 2001 ISO 12006-2: Organization of Information about Construction Works – Part 2: Framework for Classification of Information standard provided a matrix of information that is grouped into three primary categories:

- construction resources;
- construction processes; and
- construction results.

The primary categories were then divided into 15 tables for organising construction information. For the 'system' to have homogeneity and be effective in application, clearly these tables all needed to embody appropriate characteristics to facilitate cross-mapping by users. ISO 12006-2 has been under revision for some time and that status has continued into 2015. ISO 12006-3:2007 specified a language-independent information model which could be used for the development of dictionaries used to store/provide information about construction works. It enabled classification systems, information models, object models and process models to be referenced from within a common framework.

In the UK, the Common Arrangement of Work Sections (CAWS) was developed by the Construction Project Information Committee (CPIC). CAWS evolved as a result of investigations suggesting that quality of project documentation could impact significantly on the quality of construction process in terms of time/cost/quality. CAWS also provided a consistent protocol for labelling and cross-referencing between drawings, specifications and bills of quantities. It was first published in 1987 and was updated by CPIC in 1998 to align it with the Unified Classification for the Construction Industry (Uniclass).

Uniclass was the UK implementation of BS ISO 12006-2. As BS1192:2007 recommends the use of the Uniclass classification protocol, when BS1192 is referenced as a prerequisite standard within a BIM project, a knowledge of classification kicks in as a necessary prerequisite. In 2011, CPIC used the NBS proposals for reclassification of the work sections in CAWS. That work resulted in Uniclass 1.4 morphing into Uniclass 2, which offered a searchable application providing unified tables for structured information at key levels, including complexes, activities, spaces, entities, systems and work results. That iteration of Uniclass was considered to offer a more holistic perspective by including civil engineering alongside architecture and landscape, as well as more effective description of systems in performance terms, and a whole life perspective to include post-handover aspects of building and facilities management.

Uniclass 2 was the classification protocol adopted by NBS Create and designed with BIM at its centre. The application claimed to be useful as a key tool for structuring specification data within a Level 2 BIM environment. A key driver underpinning the NBS initiative was to create a single unified protocol which could allow designers to author specification information for architectural design, structural/services and external landscaping, thus enabling the entire project team to have access to a single, integrated resource.

Anders Ekholm noted that, in relation to BIM scenarios, construction 'objects' should be able to accommodate progressive levels of definition and complexity as design evolves. Ekholm also cited Charles Eastman's suggestion that the basic structure of a defined 3D object should allow successive iterations of detail, as well as the facility for deconstruction into separate component parts. Ekholm also argued the benefits of using the same semantic structure for objects across a building's whole life journey from design, through construction and post occupancy. For example, the enclosing external wall of a single family house design might start its BIM journey as a freehand line diagram on a concept sketch then evolve as nominal thickness is added at outline design stage, followed by generic wall construction in BIM authoring software. That iteration can then be progressively updated until the wall is an accurate digital representation of the real-world object defined both by appropriate geometries and embedded alphanumeric information (e.g. specification) as metadata.

Commentating on the NBS Create software at beta testing stage, reviewer Russ Green noted, 'we really like the way the sections are additive, systems can be built up, starting initially with outline specifications only linked to quite conceptual BIM objects. The model and spec develop together as the design develops' (Anon n.d.a). Classification is one of the fundamental building blocks of information management in construction. Understanding the taxonomies and nuances of classification systems is a prerequisite for meaningful engagement with Level 2 BIM.

Since the demise of CAWS, the lack of a complete, stable and universal format to serve the UK construction sector is manifest in BIM library objects offered by manufacturers typically being tagged with a range of classification descriptors (Figure 16.4), including UNSPC (United Nations), IFC, Uniclass 1.4, Uniclass 2.0 and the NBS (aligns with Uniclass 2.0) specification codes. The situation is made even more complicated by the fact that BIM authoring packages may use different classification formats as defaults. For example, Autodesk Revit uses the American standard OmniClass, which (like Uniclass 2) is aligned with ISO 12006-2.

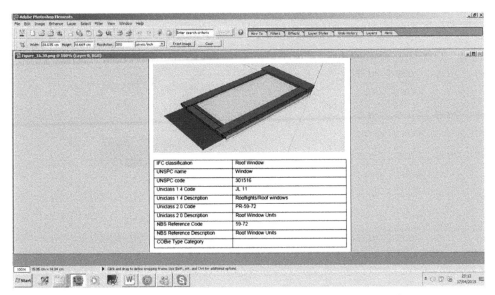

Figure 16.4 Roof window from BIMobject® library embedded with classification descriptors

As discussed in Section 15, the NBS BIM Toolkit offers an online resource, free at the point of delivery. The Toolkit also provides a structure for setting up a Level 2 BIM project and maps project information across a range of process and product models, including the RIBA Plan of Work 2013, the NRM method of measurement and classification templates. The Uniclass 2 protocol was extended, rebranded as Uniclass 15 and embedded into the application. The BIM Toolkit was designed around a unified digital structure which provided mapping and guidance so objects could be configured at a project level to have appropriate digital classification, plus step-by-step support to define, manage and validate BIM data within project-specific environments.

16.4 Populating and managing information exchanges

Architect, author and CIC BIM Regional Ambassador Steve Race used the phrase 'BIM Demystified' as the headline for his book which outlined BIM concepts, narratives and protocols. The title highlights a particular challenge for small organisations taking first steps from more conventional built environment cultures into the world of digital data and interpreting the terminology which BIM has made its own. The mystique surrounding BIM language could be interpreted by some to imply distinctiveness and the possession of special knowledge. However, many elements of contemporary BIM jargon have simply cross-pollinated from other industries. For example, whether the commonly used 'workflow', with its roots in repetitive manufacturing processes, has any place in built environment terminology is open to question, but its use is now ubiquitous in BIM circles.

Similarly 'collision avoidance' or clash detection (Figure 16.5), which is another way of describing 3D spatial dimensional co-ordination (usually between discipline-authored BIM models) has origins which reference conflicting geometries in computing technologies and gaming. Also, existing industry concepts may be tweaked and/or renamed as BIM mantras; the term 'soft landings', which morphed into 'Government Soft Landings' is an example of

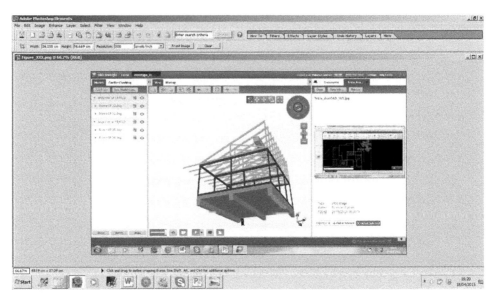

Figure 16.5 Discipline-specific models combined in Tekla BIMsight online viewer for dimensional co-ordination (clash detection)

nuances on established meanings which tend to populate BIM vocabularies. Some may find it anomalous that while BIM standards use epithets such as 'plain language questions' to imply simplicity and clarity, to the uninitiated, BIM jargon may appear unfamiliar and arcane and require some interpretation.

BIM's language needs to be learned, assimilated and applied to situations which people with industry experience will discover are familiar processes and milestones on the journey from construction project inception to completion and beyond. The term 'data drops' or in Government Task Group language 'COBie data drops' is a case in point, where tried and tested scenarios have been re-branded to incorporate BIM terminologies.

Data drops are quite simply predetermined points along a project's BIM journey at which information is pulled from the CDE to service a range of functions. While the 'data drop' terminology is a fundamental building block of the prevalent BIM lexicon, the function of staged exchanges of information has long been embedded into the processes of designing/ making/using buildings. For example, the process of extracting orthographic drawings from the model/s for a planning or building warrant submission would form part of a predetermined information exchange. If the RIBA Plan of Work 2013 process model was being used to steer a project, the content and format for BIM information exchanges would be defined at the end of each work stage.

A couple of key differences in comparing conventional with BIM practice is that data drops or staged information exchange points will have been identified within the Employer's Information Requirements (EIR) for a project where the management of digital data is defined by the PAS 1192-2 template. Also, the information exchange/s may include textual files (metadata) extracted from data-rich BIM models and/or drawings and visualisations. The format for drawing outputs would typically be in 2D PDF format, as might be required for contractual purposes (obtaining client approvals, for example), or to service applications for statutory approvals and/or construction purposes.

Typically, metadata files embedded in BIM environments are structured as spreadsheets and populated with information about spaces making up a building, room data sheets, specification data on components, assemblies, and the like. When a new or refurbished building is handed over by the design-construct team to the client, there is an immediate need for data required to service user and building maintenance requirements to be accessible, fit for purpose and up to date.

Traditionally, those functions would have been captured and served by a paper-based operations and maintenance (O&M) manual. The rationale underpinning PAS 1192-2 is that data aggregated and stored in BIM models (Project Information Models) during the design phase can be transferred to Asset Information Models to service post-occupancy requirements. The protocol which has been developed to articulate post-occupancy protocols from information management is PAS 1192-3.

Over the last few years, the Construction Operations Building information exchange (COBie) schema developed in the USA for military facilities (and available since around 2007) has been prominently publicised as industry protocol for metadata and BIM scenarios. COBie was primarily envisaged as a spreadsheet-based facilitator for gathering data on spaces, equipment and services which would be used post-handover for computer-based building maintenance. One of the perceived advantages of using COBie is that containers for metadata held in BIM models can be progressively populated with information as the design evolves from inception forward to the post-occupation phase of a building's whole life cycle. COBie files are formatted to support two distinctive situations for interacting with data requirements.

The first instance is the exchange of information with software systems which can import and use COBie data in its underlying standard format (the Industry Foundation Class [IFC] protocol). A typical application would be the import of digital data to feed proprietary open source facilities management software. For example, to service space planning requirements for an office refurbishment. Or in the situation where an unscheduled maintenance call-out was made to repair a defective heating installation, information transferred from pre- (PIM) to post-handover (AIM) BIM models could be interrogated via FM software to provide service data via a handheld device on site.

The second situation is the one which would probably have the widest possible range of applications for small practitioners, contractors, suppliers and manufacturers: it uses a ubiquitous spreadsheet format as COBie output. The spreadsheets can be populated with relevant data from BIM models and exported to a database and/or word-processing software depending on organisational requirements/capabilities. The NBS BIM Toolkit uses a variant of COBie (COBieLite) which utilises the XML schema for data management.

At first viewing, COBie output does present the unfamiliar user with a bewildering array of spreadsheet fields, implying a complexity of organisation and content. Having said that, the protocol does offer a logical and structured template for the assimilation and exchange of metadata over time. Plus, COBie is now embedded in the UK Government's suite of BIM standards which define inputs and outputs for a BIM Level 2 compliant project. Ultimately, the timing and content of information being extracted from digital environments will be project and process specific. COBie data is just one from a range of outputs which may be pulled from BIM models. Perhaps the key difference compared with more traditional methodologies for generating and handling construction data is that expectations for outputs ('deliverables' in BIM jargon) are flagged up with much greater precision from the outset, particularly when the PAS 1192-2 protocol is applied as a framework for information management.

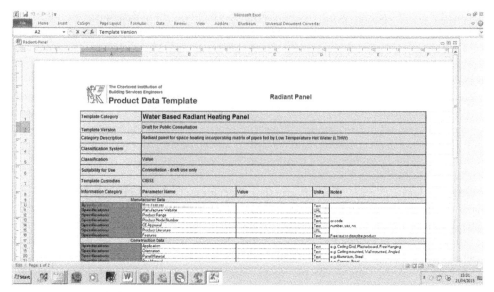

Figure 16.6 CIBSE draft generic product data template (PDT) as a medium for logging product-specific information in a BIM compliant format. Radiant heating panel example

With a significant number of manufacturers and suppliers among their member community, the Chartered Institution of Building Services Engineers (CIBSE) has been proactive in developing and testing their own COBie-compliant protocol for handling metadata: the Product Data Template (Figure 16.6). As CIBSE has argued, because the UK Government are requiring Level 2 BIM modelling by 2016, all BIM users need product data and many have already requested it from manufacturers. According to CIBSE, there is confusion as to how this objective can be achieved without much repeated, wasteful effort from companies creating customised data sets for prospective customers. The Product Data Templates (PDTs) are an evolving protocol which has been developed for the building services industry by CIBSE to address the problem by standardising metadata fields in COBie-compliant formats.

As a leading heating manufacturer, Remeha Commercial was one of the first building services product manufacturers to embed BIM into specification protocols. However, with no validated template available to act as a benchmark, Remeha created its own standard and began by carrying out research with its customers to identify and evaluate the level of information needed on their Level 2 BIM projects. Results from that scoping exercise fed into the creation of a product-specific library of BIM objects (Figure 16.7).

The Remeha BIM library consists of digital Autodesk Revit native files embedded with metadata ranging from information on the geometric configuration, size and weight of the product to the heat outputs, carbon and NO_x emissions, service and maintenance areas, and maintenance schedules. Remeha BIM files are regularly updated, providing specifiers, clients' building services engineers and information managers with accurate and reliable information which is intended to save time in product research, facilitate potential changes, promote smarter design and increase productivity. Similar to CAD blocks, the company's proprietary BIM objects can be inserted directly into the BIM-authored design environment. By providing relevant and accurate product-specific data, the accurate representations of real-world objects enable pipework runs, flues and pumps to be sized and drawn with little input from the

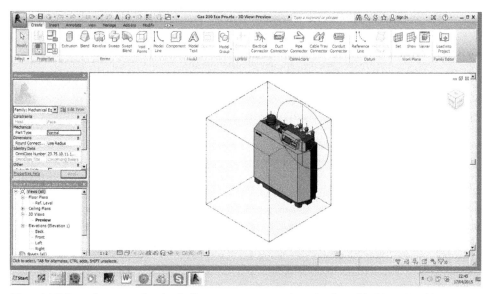

Figure 16.7 Remeha gas condensing boiler as true representation of real-world object. Combines geometry with metadata which can be extracted for installation and maintenance

design engineer. Claimed advantages include increased accuracy of representation, simplifying future maintenance and offering potential savings on time and cost aspects of installations. The geometrical accuracy and richness of embedded metadata required from specific objects will vary depending on their fit with particular stages of the design process.

In developing a BIM project, individual discipline digital models may evolve at different rates and incorporate varying levels of digital data by the time agreed points for information exchanges are reached. That situation is less of an issue than ensuring that discipline-specific information defined in the BIM execution plan is available for exchange at the appropriate project milestones and has been validated before being shared and published. Also to ensure procedures are in place so that file transfer between disciplines (for example, using the IFC protocol) will be viable and not result in significant loss of data. The idea of using open source file protocols like IFC to underpin cross-disciplinary 'dialogue' between BIM models is laudable in principle. Feedback from industry suggests that design–construct teams will resort to whatever techniques will get the job done. In practice, reliance on native file sources for information flow between BIM models may represent a more reliable methodology than using open source file exchange protocols.

16.5 Integrating inputs/outputs across disciplines

Conventional procurement methodologies still rely substantially on 2D orthographic representations and the exchange of associated documents (paper or digital) to inform the iterative processes associated with the design phases of making buildings. As drawings may typically be exchanged as PDF files supplemented by schedules and bills of quantities in similar formats, there has already been some engagement with digital environments by small design and construction organisations. With many BIM scenarios, 2D drawing output is still the norm for some 'deliverables', for example, for contracts, statutory approvals and construction purposes.

The increased use of digital tools to support project development and decision-making has triggered the onset of a new phase in the roles and relationships between construction professionals. That gradual shift in workplace practice, driven by the increased use of electronic data and other factors (for example, Government-led drivers for procurement/asset management), may impact on the nature of cross-disciplinary collaboration.

Accelerating the volume of useable digital data and the rate at which information can be exchanged within project environments has introduced new pressures for the UK construction 'industry', in particular for SMEs and micro organisations. These pressures are diverse and manifest in a number of ways. For example, in challenging the motivation for built environmental designers and constructors to engage with and manage unfamiliar digital tools, both as individuals and for cross-disciplinary dialogue within projects. Small firms will tend to equip themselves with whatever digital tools fit best with their business models and client needs. In 2011, Rizal Sebastian cited survey returns which suggested that 2D CAD is still the representational technique most commonly used for design work (over 60 per cent) while 'BIM' was used in around 20 per cent of commissions for architects and in around 10 per cent for engineers and contractors.

In 2012, the National Federation of Builders (NFB) polled its members about BIM. The response suggested that there was currently a general unwillingness to invest in and develop BIM capability despite a strong business imperative to do so. More than half of the respondents were either waiting for free training, or had no intention of up skilling to do BIM. That aspect of the data gathering was deemed important by the NFB because the data suggested the technical capacity of SMEs and micros to adopt BIM in the immediate future would likely be more limited than large contractors. In terms of engagement with construction projects utilising either 3D models, clash detection tools, schedule integration software or similar facets of using digital information, 64 per cent of respondents said they had no experience of using these techniques. Among SME contractors, that figure increased to around 70 per cent or more.

Looking through an industry lens, if the NFB scenario was representative of other construction sector SMEs, in time it may differentiate between those who do BIM (as defined by current industry drivers/standards) and those who don't, principally because it is not a priority for realising business objectives. Whether or not that premise pans out in practice remains to be seen; other paradigms for engaging with/managing digital data may evolve. At the point when the UK BIM Task Group's strategy document Digital Built Britain was published in February 2015, there was no compelling evidence in the frame to suggest that construction UK was tooled up to address BIM Level 2, let alone progress to the next 'level'. In that context, the industry priority for SMEs and micro organisations is to step up to the mark and engage with digital tools and methodologies.

Whatever the size of project, collaborative working using BIM requires a co-ordinating role fulfilled by a design-construct team member with information technology skills as well as construction process experience and insight. That person needs to be able to manipulate data and software (hard skills) as well as interact with design-construct team members (soft skills), develop technological solutions required for BIM functionalities and manage the flow of digital information across disciplines during project development and execution.

The role has variously been described as 'BIM Manager', 'Information Manager', BIM Co-ordinator' and 'Project Information Manager'. As has been discussed, smaller projects will seldom justify a separate appointment to enact the function of co-ordinating BIM scenarios, and the key task of information management will need to be assigned to an appointee from within the design-construct team. Potentially the BIM co-ordinating function could be delivered by a team member from a range of construction-related backgrounds.

Commenting on the role of the information manager, Steve Faulkner of structural engineers Elliot Wood said:

> On smaller projects or where teams and individuals are new to BIM it can be quite daunting, and indeed challenging, having to trawl through complicated BIM documents trying to find the relevant parts. BIM documentation needs to be simple and concentrate on the key features. We should only use the parts of PAS 1192-2 that will benefit the project. Ten pages of important information are likely to be read, 200 pages of overly detailed and complex information will not, and sometimes the important bits will be lost in the process.
>
> (Faulkner 2015)

From the outset of a new project, a substantive discussion needs to be developed with the client about the extent to which digital data and BIM will be developed and utilised, and about the extent to which data-rich models may be incorporated, referenced and can be relied upon to provide validated and reliable data to serve a range of project needs. These requirements need to be defined at the stage from which project aims, objectives and staged outputs are defined (typically RIBA Stage 0, 'Strategic Definition') and embedded in documentation for information management.

Through this formative period, the design team begin their project journey with a common understanding of how digital data and BIM will be deployed across the project timeline, in particular during pre- and post-contract phases of execution. The designated 'information manager' (CIC terminology) has a pivotal role to play from that point on. Key tasks include:

1. Articulating the client's information requirements in terms of data inputs/outputs required for design, construct and post-occupancy purposes. Embedding that data into the Employer's Information Requirements (EIR) document. Establishing most effective protocols for client engagement with the design process to facilitate decision-making.
2. Defining the project data management structure, how that will be developed through discipline-specific BIM models and how these models will be co-ordinated, for example through a federated model structure.
3. Agreeing protocols for model ownership, referencing and identifying any intellectual property issues which may need to be formalised in contractual arrangements. For example through use of the CIC BIM Protocol.
4. Developing a BIM Execution Plan (BEP). CPIC offer downloadable pro formas (CPIx) for both pre- and post-contract purposes. The CPIx template embodies examples for BEP and Project Implementation Plan (PIP). Cross-referencing between BEP and project process model, for example RIBA Plan of Work 2013, so that appropriate milestones for data inputs/outputs can be identified and embodied in the BEP.
5. Embedding relevant reference standards for information management into the BEP. For example, BS1192:2007 and PAS 1192-2. Establishing templates for file management/sharing, quality control protocols, dimensional co-ordination and processes for referencing, checking and validation of models as design evolves.
6. Defining required levels of model development and 'deliverables' at each work stage, for example required outputs to service cost/project planning and obtaining statutory approvals.

7. Setting up a CDE and briefing design team members on procedures for engaging with shared model environments where these are used. For example the use of proprietary model viewers for checking dimensional accuracy and fit across discipline models.
8. Incorporating procedures for managing digital data within the CDE as the project evolves. Maintaining agreed quality control standards, for example file updating and backup at regular intervals.
9. Managing information flows as required for digital prototyping and trial manufacturing purposes. For example predictive energy performance analysis, virtual builds to test general arrangement and fit of components.
10. Troubleshooting and resolving communication errors in relation to effective running of the CDE as an environment for information exchange to inform collective decision-making within the design/construct team.
11. Managing the capture and flow of data required to serve the client's post-occupancy needs (as defined within the EIR). Co-ordinating that data into useable form for the client at handover. Depending on the project size/type, data formats for post-occupancy purposes could range from folder structure containing spreadsheets, manufacturer's data and the like for a small one-off project to automated FM software for larger more complex projects or, for example, where client needs determine that a property portfolio is managed centrally.
12. Demonstrating effective team building communication and leadership skills for information management as applied to the attainment of strategic and operational project goals/outcomes.

These tasks have been set out in an indicative chronological order of project evolution for guidance, but no hierarchy is implied. Perhaps the most important attribute for an information manager has been left until last. As well as administering the practical aspects of information flow, commanding the trust and respect of the design-construct team is a key and common attribute of both design and information manager roles. In practice, the design/information management functions may overlap significantly and in some instances may merge into a combined role, particularly for the management of small projects.

The enactment of historical discipline roles is less important with BIM than the requirement to clarify and record individual and collective ownerships for inputs/outputs of digital data from project inception. In that context, all disciplines need to be prepared to take on some level of responsibility as information managers when using digital tools. The imperative needs to be incorporated into education and training, both within and across built environment disciplines to have some meaningful impact on the future of the construction industry in the UK. The challenge of integrating discipline inputs into the processes of making and managing buildings is best addressed at undergraduate level, before the die is cast and established discipline role models are perpetuated for another generation.

16.6 BIM templates for virtual construction planning and building for real

Apart from assimilating new knowledge about BIM procedures and transferring familiar skills from conventional to BIM scenarios, something else SMEs and micros need to consider is looking beyond familiar and sometimes comfortable templates for design, cost planning, scheduling and construction, and thinking more radically about how they use data in attaining business-driven outcomes by pushing the boundaries (even just a little). That exercise needs to be developed across the full spectrum of an organisation's activities.

Potentially, it is a win-win situation, because an internal review of how an SME or micro organisation uses data (forget the term 'BIM' in carrying out that exercise) could either feed into and consolidate existing processes or assist an organisation to develop new (and possibly more efficient) workflows. New ways of executing familiar processes may emerge as a consequence. The UK construction industry is burdened with a significant track record (perhaps, more accurately, reputation) of being starved of research and development resources. The usual suspect in the blame game is lack of readily available funding to do R&D. Perhaps that perspective belies a deeper cultural malaise with roots in inherent resistance to change.

It is that last point which may be crucial in overcoming resistance to developing a competitive edge in today's data-driven society. Why should construction professionals exclude themselves from new workplace methodologies like BIM and continue to do things the same old way? US consultancy Kieran Timberlake combines mainstream architectural practice with development of applications, including software, which the company claims can 'conduct iterative and highly accurate life-cycle assessments on any scale, from component to whole building, as an integrated component of the BIM design workflow' (Ravenscroft 2015). That kind of versatility is driven by an holistic and strategic perspective of data in use, combined with the aspiration to maintain a competitive edge in the industry.

Contemporary BIM discussions generally are fraught with complexity, contradiction and sometimes polarised viewpoints. As has been discussed, there is no single, clear and unequivocal way of defining or doing BIM. UK construction lacks a body of knowledge which is evidence based and has drawn data from a significant portfolio of small BIM projects developed in the field. Using a facilities management metaphor, BIM UK needs to be informed by a feedback/feedforward process which can, over time, develop reliable templates for 'best practice'.

These machinations can be confusing and even disorientating for small organisations with aspirations to take 'first steps' away from conventional practice and towards information-based scenarios. As Barry Miller noted, one of the prevalent idiosyncrasies with BIM is that it attempts to seduce with promises of increased productivity, fewer disruptions, less duplication, reduced claims and greater profits. Yet, however worthy the efforts of SMEs and micros may be in developing templates for lonely BIM, in a collaborative situation, the willingness and effort needed by all participants to fully realise these perceived advantages can't be guaranteed.

Like most software, the latent power of BIM authoring packages as relational databases tends to be untapped in practice, particularly if the mindset and practice leans towards using conventional 2D output for contract and construction purposes. BAM Construct's digital construction manager Mark Taylor described his company's mentoring of their partition suppliers in using BIM tools to develop templates for estimating, scheduling and cost purposes. But that example is still a long way from representing a norm in UK construction, particularly within supply chains. The potential to unlock and deploy data remains significantly underutilised. Even in the transition from conventional to BIM tools, the default remains 2D output. It would be a real paradigm shift to see more evidence of new work practices which could link design and construction planning with manufacturing from within digital environments.

Technology's pace of change is relentless. There is a lingering gut feeling among some that the aspiration to reach out towards ubiquitous and ultimately fully collaborative data-rich models (Level 3 BIM) may be overtaken by the increased use of cloud-based applications which have access to huge data sets, like the Loughborough University example, or by the deployment of other digitally driven and enabling techniques, like augmented reality (AR) and the Internet of things, to do specific jobs. These could range from tasks related to the daily grind of project evolution in getting buildings through the design process and onto site, to longer ranging/holistic whole life objectives, such as energy management and environmental impact scenarios. What Chairman of HM Government BIM Working Group Mark Bew described as the 'Facebook Revolution' in relation to Level 3 aspirations may already be on its way for UK construction, but via a different flight path than has been envisaged by UK BIM strategists.

One challenge is that while the industry struggles to engage with Level 2 BIM, the policymakers are simultaneously raising the bar. However, as the use of 'big data' and powerful predictive data analytics techniques becomes more and more accessible to small organisations, the Level 3 mantra may dissipate and be subsumed into new techniques for data management. As software developer Pat Hanrahan said, 'the reason big data is impacting every one of us is the data oozing out of everything ... it's like electricity flowing throughout an organization – everyone can tap into it on command to answer the individual questions their jobs demand' (Woods 2011). Data is something everyone in an organisation receives, and they should be able to use it immediately to make timely decisions.

The Tableau software being used by project management consultants Turner and Townsend offers a taster of how 'big data' applications may filter down from urban scale to project-specific frames of reference by allowing built environment disciplines to see, understand and apply data at project level through alternative formats to databases and spreadsheets. Users can capitalise on the functionality of being able to query large data sets from a variety of sources in an ad hoc manner and drag and drop them into the proprietary application to be instantly converted into graphical images. That fast and analytical use of data enables users to visually see patterns and identify trends and insights quickly and efficiently.

Connecting to a data source or blending from multiple data sources is virtually instantaneous and can generate visual trends and patterns within minutes. Also, the software's mapping function allows the visualisation of cost, schedule, risk and other data types geographically. For example, when 2014 UK flooding caused water-handling systems to overload and generate unusually high maintenance requirements, visualising sites geographically and comparing with historical trends enabled the impact of the abnormal weather conditions to be identified and mitigated. At project level, viewing a construction site on a digital map which can pull and filter from 'big data' sources allows users to interact with potential conflicts which could result in project delays, road improvements and other potentially disruptive factors. It is only a matter of time before these 'big data'-driven tools scale down and become affordable for small organisations.

Bibliography

Alhava, O., Laine, E. and Kiviniemi, A. (2015) 'Intensive Big Room Process for Co-Creating Value in Legacy Construction Projects', *ITcon*, 20, [online]. www.itcon.org/cgi-bin/works/Show?2015_11 [accessed 27 February 2015].

Anon (1996) *BS 7000-4:1996 Design Management Systems. Guide to Managing Design in Construction*, BSI Standards Ltd.

Anon (2001) *ISO 12006-2:2001 Building Construction – Organization of Information about Construction Works – Part 2: Framework for Classification of Information*, ISO, [online]. https://www.iso.org/obp/ui/#iso:std:iso:12006:-2:ed-1:v1:en [accessed 27 February 2015].

Anon (2002) 'Building Information Modelling', Autodesk White Paper, Autodesk, [online]. www.lai-serin.com/features/bim/autodesk_bim.pdf [accessed 24 March 2015].

Anon (2007) *ISO 12006-3:2007 Building Construction – Organization of Information about Construction Works – Part 3: Framework for Object-oriented Information*, ISO, [online]. www.iso.org/iso/catalogue_detail.htm?csnumber=38706 [accessed 2 March 2015].

Anon (2008) *BS 7000-1:2008 Design Management Systems. Part 1: Guide to Managing Innovation*, BSI Standards Ltd, [online]. http://shop.bsigroup.com/Browse-by-Sector/Design/BS-7000-Series–Design-Management-Systems/ [accessed 27 February 2015].

Anon (2009) *Building Information Modelling (BIM) Guidelines and Standards for Architects and Engineers*, Division of Facilities Development, Department of Administration, State of Wisconsin, 1 July, [online]. ftp://doaftp1380.wi.gov/master_spec/BIM%20Guidelines%20&%20Standards/BIM_Guidelines_and_Standards%206-09.pdf [accessed 18 February 2015].

Anon (2012) *AEC(UK) BIM Protocol, Version 2*, AEC UK, [online]. https://aecuk.files.wordpress.com/2012/09/aecukbimprotocol-v2-0.pdf [accessed 2 March 2015].

Anon (2013a) *Guide, Instructions and Commentary to the 2013 AIA Digital Practice Documents*, American Institute of Architects, [online]. www.aia.org/groups/aia/documents/pdf/aiab095711.pdf [accessed 25 February 2015].

Anon (2013b) *BS 7000-4:2013 Design Management Systems. Part 4: Guide to Managing Design in Construction*, BSI Standards Ltd.

Anon (2013c) COBie UK 2012, Building Information Modelling (BIM) Task Group, [online]. www.bimtaskgroup.org/cobie-uk-2012/ [accessed 3 March 2015].

Anon (2013d) *Pre-Contract Building Information Modelling (BIM) Execution Plan (BEP)*, CPIx on Line, V2.0, CPIC, [online]. www.cpic.org.uk/wp-content/uploads/2013/06/cpix_pre-contract_bim_execution_plan_bep_v2.0.pdf [accessed 25 February 2015].

Anon (2014a) 'Common Arrangement of Work Sections', *Designing Buildings Wiki*, 5 December, [online]. www.designingbuildings.co.uk/wiki/Common_Arrangement_of_Work_Sections [accessed 27 February 2015].

Anon (2014b) 'Industry Debate Puts Opportunities of Collaboration in Implementing BIM in the Spotlight', *News*, Chartered Institution of Building Services Engineers, 28 October, [online]. www.cibse.org/news/october-2014/industry-debate-puts-opportunities-of-collaboration [accessed 16 April 2015].

Anon (2014c) 'NBS Live – Introducing the BIM Toolkit', NBS, [online]. www.thenbs.com/topics/bim/articles/NBS-live-introducing-the-BIM-toolkit.asp [accessed 3 March 2015].

Anon (2014d) 'Preparing for BIM', *CIBSE Journal*, Chartered Institution of Building Services Engineers (CIBSE), June, [online]. www.cibsejournal.com/archive/PDFs/CIBSE-Supplement-2014-06.pdf [accessed 16 April 2015].

Anon (2015a) 'Innovative App Unveiled to Help Tackle Information Overload in 3D Virtual Building Technology', *Latest News from Loughborough University*, Loughborough University, 15 April, [online]. www.lboro.ac.uk/news-events/news/2015/april/3d-information-retrieval-app.html [accessed 16 April 2015].

Anon (2015b) 'Reports', *Building Information Modelling*, NBS, [online]. www.thenbs.com/topics/bim/reports/index.asp [accessed 18 April 2015].

Anon (n.d.a) 'Origin3 Studio', Beta Test Programme, NBS, [online]. www.thenbs.com/support/betatesting/Russ-Green-Origin3Studio.asp [accessed 25 February 2015].

Anon (n.d.b) 'Product Data Templates', Chartered Institution of Building Services Engineers, [online]. www.cibse.org/knowledge/bim-building-information-modelling/product-data-templates [accessed 24 March 2015].

Banks, J. (2013) 'There are Four BIM Flavors', *Shoegnome*, [online]. www.shoegnome.com/2013/01/31/there-are-four-bim-flavors/ [accessed 15 February 2015].

Barnes, P. and Davies, N. (2014) *BIM in Principle and in Practice*. London: Institution of Civil Engineers, ICE Publishing.

Barraclough, A. (2015) 'How Can Architects Make Contractors Deliver their Vision?', *bdonline*, 26 February, [online]. www.bdonline.co.uk/bim/how-can-architects-make-contractors-deliver-their-vision?/5074052.article [subscription] [accessed 27 February 2015].

Benson, S. (2010) 'From Lonely BIM to Social BIM: Moving Beyond Design to FM', *Archibus White Paper*, Archibus, [online]. www.informi.dk/fileadmin/UploadInformi/PDF/Aktiviteter_presentations/FM_til_hospitaler/WhitepaperBIM.pdf [accessed 15 February 2015].

Bogen, C. and East, B. (n.d.) 'COBieLite: A Lightweight XML Format for COBie Data', National Institute of Building Sciences, [online]. www.nibs.org/?page=bsa_cobielite [accessed 3 March 2015].

Cousins, S. (2014) 'Turner and Townsend Turns to Tableau to Manage Data', *BIM+*, CIOB, 5 November, [online]. http://bim.construction-manager.co.uk/news/turner-townsend-turns-tableau-manage-data/ [accessed 9 March 2015].

(2015) 'BAM Construct's Formula for Level 2', *BIM+*, CIOB, 6 March, [online]. http://bim.construction-manager.co.uk/people/bam-constructs-formula-level-3/ [accessed 9 March 2015].

East, B. (2014) 'Construction-Operations Building Information Exchange (COBie)', *Whole Building Design Guide*, a program of the National Institute of Building Sciences, [online]. www.wbdg.org/resources/cobie.php [accessed 3 March 2015].

Eastman, C., Fisher, D., Lafue, G., Lividini, J., Stoker, D. and Yessios, C. (1974) 'An Outline of the Building Description System', *Institute of Physical Planning Research Report No. 50*, September, Carnegie-Mellon University, [online]. http://files.eric.ed.gov/fulltext/ED113833.pdf [accessed 26 March 2015].

Eastman, C., Teicholz, P., Sacks, R. and Liston, K. (2011) *BIM Handbook: A Guide to Building Information Modeling for Owners, Managers, Designers, Engineers and Contractors*, USA: John Wiley and Sons.

Ekholm, A. (1996) 'A Conceptual Framework for Classification of Construction Works', *ITcon*, 1, [online]. http://itcon.org/cgi-bin/works/Show?1996_2 [accessed 26 February 2015].

Ekholm, A. and Haggstrom, L. (2011) 'Building Classification for BIM Reconsidering the Framework', *Proceedings of CIB W78-W102 2011: International Conference*, Sophia Antipolis, France, [online]. https://lup.lub.lu.se/search/publication/2201252 [accessed 26 February 2015].

Eynon, J. (2013) *The Design Manager's Handbook*. London: CIOB, Wiley-Blackwell.

Faulkner, S. (2015) 'The Importance of an Information Manager', *BIM+*, CIOB, 13 February, [online]. http://bim.construction-manager.co.uk/people/importa5nce-inf5ormation-mana7ger/ [accessed 9 March 2015].

George, D. (n.d.) 'BBC Media City, Salford Quays', Capita Property and Infrastructure, [online]. www.capitasymonds.co.uk/projects/all_projects/bbc_media_city,_salford_quays.aspx [accessed 25 February 2015].

Jernigan, F. (2007) *BIG BIM little bim*, USA: 4 Site Press.

Knutt, E. (2015a) 'Level 3 BIM Culture Change is Our "Facebook" Revolution Says Bew', *BIM+*, CIOB, 5 March, [online]. http://bim.construction-manager.co.uk/news/facebook-revolution-says-bew/ [accessed 9 March 2015].

(2015b) 'Satellite BIM Platforms on their Way for Tier 2 Contractors', *BIM+*, CIOB, 12 April, [online]. www.bimplus.co.uk/news/satellite-bim5-platf4orms-t6heir-way-tier-2/ [accessed 15 April 2015].

(2015c) 'Mace Adopts SLIMBIM2GO For BIM "Cartoon" in a User Friendly App', *BIM+*, CIOB, 26 April, [online]. www.bimplus.co.uk/technology/user-friendly-tablet-app/?utm_source=newsletter&utm_medium=email&utm_campaign=bim_2015-03-11 [accessed 27 April 2015].

Miller, B. (2014) 'BIM's Potential Has Yet to Be Realized', Tech Trends, *Construction Executive*, 19 February, [online]. http://enewsletters.constructionexec.com/techtrends/2014/02/bims-potential-has-yet-to-be-realized/ [accessed 9 March 2015].

Napier, B. (2008) 'Wisconsin Leads by Example', *Journal of Building Information Modelling*, Fall, [online]. ftp://doaftp1380.wi.gov/master_spec/BIM%20Guidelines%20&%20Standards/BIM%20Article%202008%20-%20Wisconsin%20Leads%20by%20Example.pdf [accessed 18 February 2015].

NFB (2012) *BIM: Ready or not?*, National Federation of Builders, [online]. www.builders.org.uk/resources/nfb/000/318/333/NFB_BIM_Survey_BIM-ready_or_not.pdf [accessed 25 February 2015].

Race, S. (2012) *BIM Demystified*. London: RIBA Publishing.

Ravenscroft, T. (2015) 'UK Version of Kieran Timberlake's Eco App Tally Could Be on its Way', *BIM+*, CIOB, 6 March, [online]. http://bim.construction-manager.co.uk/technology/eco-app-could-be-its-way/ [accessed 9 March 2015].

Sebastian, R. (2011) 'Changing Roles of the Clients, Architects and Contractors through BIM', *Engineering, Construction and Architectural Management*, 18, [online]. www.emeraldinsight.com/doi/abs/10.1108/09699981111111148 [accessed 25 February 2015].

Tomaszewski, L. (2014) *BIM From the Architect's Perspective*, Chapman Taylor. [online] www.bpcc.org.pl/att/3e3ec1f0-53c5-4a41-b412-477e85d1ec26_bim_chapman_taylor.pdf [accessed 16 February 2015].

Van, J. (2008) 'AGC BIMForum Review', *All Things BIM*, 8 October, [online]. www.allthingsbim.com/2008_10_01_archive.html [accessed 17 February 2015].

van Nederveen, G.A. and Tolman, F.P. (1992) 'Modelling Multiple Views on Buildings', *Automation in Construction*, 1(3), December, [online]. www.sciencedirect.com/science/article/pii/092658059290014B [accessed 24 March 2015].

Woods, D. (2011) 'Tableau Software's Pat Hanrahan on "What Is a Data Scientist?" ', *Forbes*, 30 November, [online]. www.forbes.com/sites/danwoods/2011/11/30/tableau-softwares-pat-hanrahan-on-what-is-a-data-scientist/ [accessed 10 March 2015].

17 Developing strategies for the post-occupancy phase

17.1 Big pictures, soft landings and the BSRIA five-stage process

A key driver for the UK Government's BIM strategy has been its intention to shape and create an efficient, fit-for-purpose and 'sustainable' estate of public buildings and facilities. Enactment of that vison is bound up in the industry reference standards produced to date, principally the PAS 1192 suite of documents. UK construction is constantly being reminded that the PAS suite offers 'must-have' guidance for doing Level 2 BIM.

Since 2010, although central government's physical estate has shrunk by around 20 per cent (from 10.5 to 8.5 million square metres internal floor area), the executive agencies still retain a huge property portfolio. Similarly, in America the General Services Administration (GSA) is one of the largest building owners in the country and holds a massive estate of physical assets designed for 50+ years of use.

From 2003 onwards, the GSA pioneered and developed the National 3D-4D-BIM Program. A key component of that initiative was to ensure that the appropriate BIM deliverables were in place to address post-occupancy facilities management needs. The GSA's strategic rationale in linking BIM with FM was to facilitate the leverage of whole life data to provide 'safe, healthy, effective and efficient work environments for our clients'. At an operational level, the view was that BIM data generated during design and construction phases could be usefully deployed and progressively updated across the whole life cycle of buildings.

The perception was that BIM could also act as a touchstone for realisation of efficiency gains, for example, through being able to offer accurate as-built information to reduce the cost and time required for alteration/adaption, and through increasing customer satisfaction and optimising the operation/maintenance of buildings and their systems to reduce energy use. With the USA buying into the 1997 Kyoto Protocol for global CO_2 reduction (although not committing to legal ratification in 2014), it could be argued that linking BIM with built environment post-occupancy performance gains offered potential to add value across a broad spectrum. In fact, perspectives ranging from single buildings through local, regional and national macro-urban levels to contexts ring-fenced by global domains. In evaluating operational effectiveness of the premise, there is a catch-all assumption that appropriate metrics are in place to test/measure performance in use at every level.

The UK Government is also actively promoting its The Way We Work (TW3) initiative, a cross-departmental programme designed to help realise the Civil Service Reform Plan's aim of 'creating a decent working environment for all staff, with modern workplaces enabling flexible working, substantially improving IT tools and streamlining security requirements' (Anon 2014). As a first step, and to further reduce the need for office space, the plan was to drive a smarter working revolution to transform how and where civil servants work. The big targets are to 'increase productivity, reduce costs, improve well-being, and contribute to wider objectives such as localism, sustainability, and reducing pressure on the transport system'. These considerations dovetail quite neatly into BIM and FM scenarios and highlight caveats which bubble to the surface in considering buildings as 'assets' as though they were entities in some way detached from the people who use them.

While the Government BIM Task Group has been actively promoting the Government Soft Landings (GSL) message as a subset of its BIM mantra, the concept of soft landings has a long tradition dating back to architect Mark Wray's dialogue with David Adamson, Director of Estates at Cambridge University. The thrust of these interactions was to develop an innovative and holistic approach, together with practical guidance, designed to close perceived gaps between client/user expectations and building performance in use.

As the UK tradition was articulated:

> the rigid separation between construction and operation means that many buildings are handed over in a state of poor operational readiness and suffer a hard landing, particularly – as often happens – when delays have led to the telescoping of the commissioning period. Problems can be worst where complicated or unfamiliar techniques and technologies are used and nobody can understand why, or what they need to do. If the problems are not dealt with rapidly, occupants' initial enthusiasm can easily turn into disappointment.
>
> (Way and Bordass 2014)

The practicality of embedding soft landings into design/construct/post-handover was set out in a five-point series of actions tabled by BSRIA and is paraphrased below:

- inception and briefing to clarify the duties of members of the client, design and building teams during critical stages, and help set and manage expectations for performance in use;
- design development/review to incorporate more rigorous procedures established in the briefing stage for reviewing predicted performance against the original expectations and achievement of specific outcomes;
- pre-handover to have greater involvement of designers, builders, operators and commissioning/controls specialists, to strengthen the operational readiness of the building;
- initial aftercare during the users' settling-in period, with a resident representative or team on site to help pass on knowledge, respond to queries and react to problems;
- extended aftercare in years 1–3 following handover, with periodic monitoring and review of building performance.

The soft landings framework aligns with the RIBA Plan of Work 2013 and, in its seminal document 'The Soft Landings Framework', BSRIA produced a workflow diagram which maps soft landings progression and practical actions against RIBA work stages. For project-specific use, that paradigm can usefully be embedded into BIM templates along with BS1192:2007, the PAS 1192 suite and the CIC BIM Protocol. Also, relevant soft landings actions at the various work stages can be transposed into the overall frame of reference for design/project management, for example, data management, validation and appropriate inputs/outputs.

For example, in the case of a housing association new build project of 100 houses, metadata attached to a heating boiler BIM object embedded in an M+E model at LOD 500 could be audited to ascertain if operating instructions were consistent with tenant feedback incorporated into the soft landings protocol developed earlier in the design process.

The BSRIA soft landings principles are not building size or type dependent; they can be applied to any project at any scale, from a single family house to the 100-house development referred to. In that context, there is one caveat in relation to repetition: operational shortcomings in a single instance which are not intercepted and resolved during the design process are likely to be repeated over and over again during service.

That instance alone represents a powerful argument for use of the BSRIA methodology as an analytical and data-gathering tool for prototyping interactions between people, product and process in digital environments. Plus, with reference to the BIM Task Group's GSL protocol, there is sometimes little point in trying to reinvent the wheel. Also, as was noted elsewhere in relation to PAS 1192-3, making distinctions between post-occupation 'asset management' and 'facilities management' might seem a tad artificial for small property holding organisations, where the two terms effectively mean one and the same thing in practice.

17.2 The fit of post-occupancy considerations with information flow

The 2011 Building Information Modelling (BIM) Working Party Strategy Paper set out a plan to deliver a structured capability for increased BIM take-up as part of a joined-up thinking process to improve the performance of the government estate in terms of its cost, value and carbon impacts. Reflecting on the role of central governments as change agents, three reference points for BIM, small organisations and post-handover aspects of building management spring to mind:

- first, as a general but all pervasive question, whether there is broad compatibility (or not) between the Government's BIM paradigms and the ways in which small business is structured and operates;
- second (in the specific context of digital data management), whether the implied continuity between design/construct/use is of real relevance to the majority of SMEs and micros (particularly in situations where new buildings and/or alterations/adaptions are simply handed over to clients and design-construct teams move on to the next project);
- third, whether BIM as we know it (based on the use of 3D data-rich models) will be subsumed into the world of big data and apps which may ultimately prove more agile/useful for small businesses and their clients in responding to soft and hard FM requirements.

As Martyn Freeman of FM consultants Mitie commented:

> if there's one expression guaranteed to cause most eyelids to droop it's 'big data'. But you ignore it at your peril, because more and more of the decisions we make on just about every aspect of work are affected by, or rely on, hard numbers. Every day we're assaulted with information, yet while we may be drowning in data, too little of it is used effectively.
>
> (Freeman 2013)

Freeman continued to reflect that:

- surprisingly few organisations can access a single view of all the minute details that impact on the quality and cost of long-term occupancy decisions;
- every aspect of a building's life is capable of being captured; not just rent, business rates and utility costs, but also usage information on a vast range of factors that can determine how effectively a building is being used;

- capturing this data and turning it into actionable information is a key part of effective facilities management, as it supports well-informed decisions which will help drive down space related operating costs;
- probably fewer than a quarter of all the organisations we meet have put in place Computer Aided Facilities Management software (CAFM) to capture and analyse this key data;
- much to our surprise, it transpired that for a lot of businesses, FM is seen as a commodity service to be delivered at the lowest cost. Only rarely does it have a strategic influence on business development.

Data needs to be captured, structured, stored and kept up to date during the design and development phases with new build developments. In the case of a refurbishment project, data sets could include the output from high-definition 3D scanners which can be exported into BIM authoring software to form part of a Project Information Model (as defined within PAS 1192-2). As handover of a building from the design-construct team to the client approaches, post-occupancy information requirements as agreed and embedded in the project EIR need to be addressed and actioned by transferring data from the Project Information Model (PIM) to the Asset Information Model (AIM) and preparing the specified portfolio of deliverables for post-occupancy use. PAS 1192-2 provides guidance but distinguishes between 'asset' and 'facilities' management, a distinction which may in practice be somewhat academic for small projects.

17.3 Post occupation and initial aftercare

The period following handover of a completed commission to the client marks a critical phase of project evolution. When responsibility for the building is transferred from the design-construct team, the client (and/or client team) will typically be presented with an operations and maintenance (O&M) manual. In some instances, that aftercare 'handbook' would comprise a single, logically structured and comprehensive document. In other situations, the client team might be handed a basic package of information referencing the various installations and equipment required to keep a building functional during use.

Initial aftercare places a wide spectrum of burdens on building owners. Some of these have legal implications, for example, compliance with statutory requirements for health and safety. Other client obligations are more people-related, in terms of the way they interact with and respond to buildings and their surrounding environments. BIM offers a huge opportunity for data required for aftercare to be generated and stored as design progresses in formats which can be readily accessed post building occupancy.

Capitalising on that process should reduce post-handover uncertainty generally. In the specific situation where a BIM project has been developed using the PAS 1192-2 protocol, the accumulation of structured data should increase the potential for a comprehensive information pool to be available whatever the size of the project. Typically, it will only be the formats for accessing and managing data which may differ. For example, small one-off projects will require data for building management, but may not require access to the sophisticated FM packages typically used to manage large facilities and property portfolios.

With some building types, such as housing, user needs (and sometimes frustrations) in the early months post handover can put pressures on the ability of support systems to be responsive and cope with demand. For example, small things like the operation of central heating controls, which seem quite straightforward to an installer, can be perceived as complex and difficult to understand by building users. In other situations, post-occupancy challenges can be more severe and lead to building closure, as in the case of the leisure centre outside Edinburgh which was shut by the local authority early in 2015 due to safety concerns less than a month after opening. Application of the soft landings process to strategic planning and project execution is likely to ameliorate these pressures and in some cases may eliminate them altogether. The 'prevention being better than cure' epithet applies aptly to embodying soft landings principles in BIM processes.

In the broad context of facilities management, the UK Government 2016 deadline and BIM, Rob Jackson of Bond Bryan Architects noted that three information deliverables would be required for publicly procured built environment assets: native models, PDF and COBie.

As discussed earlier, COBie is an open standard intended to deliver data necessary for the successful maintenance and management of buildings and infrastructure. Information embedded in COBie templates can be used for a variety of purposes. These range from 'passive' data, which is simply held in spreadsheet format in a database for record purposes, to 'active' data, which can be deployed to feed specialist software used for post-occupation building management/maintenance purposes serving 'hard' and 'soft' FM functions. The various typologies include Computer-Aided Facilities Management (CAFM) systems, Computerised Maintenance Management Systems (CMMS) and Integrated Workplace Management of Systems (IWMS). Specialist software packages are able to import COBie data and populate their own data fields. Potentially, this 'automation' of process can save building and facilities managers the time spent in manually entering the information using conventional data-entry methods.

Whether or not the full potential scope of outputs from BIM models will impact on the smaller projects typically delivered by SMEs and micro organisations had not been extensively tested at the time of writing. In particular, the fit of the COBie protocol with the operational needs of small organisations remains open to question until hard evidence from the field suggests otherwise. Having said that, one advantage of COBie is that it is able to handle data from tagged objects in a BIM authoring package. Geometric and metadata can be updated and refined through project stages as spaces, assemblies or components assume progressively increased definition during the design process.

While the post-occupancy management of smaller 'one-off' projects may not justify the use of specialist facilities management software, the principle of being able to pull data from the modelling environment and apply to service post-handover requirements applies universally, whatever the project type, size and value. Once that rule has been established, the actions of commissioning organisations and clients in defining their information needs at an early stage of project development forms a key part of developing an information strategy. That process feeds into development of the EIR document as a robust and accurate statement of data requirements.

Ashley Beighton of the Clarkson Alliance Limited (Beighton 2013) described how the 12-house, £1.4 million Meadow Road sheltered housing scheme for Worthing Homes in West Sussex had been set up as a collaborative BIM project. A key objective was to understand the changes in process and behaviours needed to work in a Level 2 BIM environment where a range of authoring tools was being used by the project team to import/export data to/from a single, federated BIM model managed by Kier's BimXtra (now Clearbox) collaborative software.

The project team decided to focus on the Asset Information Model (the information needed to efficiently and effectively operate and maintain the finished building) early on in the process, with the client's requirements then being written into the EIR. The decision was made not to embody COBie as a deliverable and instead construct an AIM which would provide the client with the following outputs on handover:

- health and safety file containing 'passive' project data;
- federated model in IFC file format containing 'passive' system data;
- spreadsheet/s containing 'active' system data.

The team's intention was that the passive data sets would effectively form record documentation for the project, while the active data would feed directly into the client's planned preventative maintenance system.

Typically BIM objects which would be transferred from a PIM to an AIM would be embedded in:

- mechanical and electrical systems;
- lighting installations, fixtures and equipment;
- plumbing systems and associated equipment;
- furniture, manufacturers and specification;
- fire protection systems;
- security installations; and
- specialist installations (for example, communications equipment).

Using FM terminology, so-called 'soft' services requiring data input to populate an AIM would include cleaning, recycling and disposal of waste, security, health and safety, furniture and equipment, continuity planning, space allocation, churn, and post-occupancy evaluation. Similarly 'hard' post-occupancy services would cover building maintenance (both planned and reactive), energy and water management, building management systems, heating, ventilation and other services installations. In addition, asset and whole life management of the project would include post-handover planning, management and control of the building over its lifespan.

Perhaps the key difference between conventional O&M and BIM approaches to post-handover processes is the early involvement of the client in the process of determining and shaping the digital data sets which will be required for effective building use, management and maintenance. That is a straightforward principle and applies equally to both small, one-off buildings and larger, more complex developments which might be contained within extensive property portfolios held by large organisations. The process can be framed and managed using the PAS 1192-2 standard. Validating the way/s in which data is stored, maintained and protected is key to informing post-occupancy activities.

17.4 Embedding information strategies for extended operations

A wide range of software applications is commercially available for the management of digital data during the occupation phase of a building's whole life use. In the case of a BIM project, depending on the building type and client perspectives, the useful life for the Asset Information Model (AIM) generated from the Project Information Model (PIM) could vary. For example, in the case of an office building which is sold on, has a change of use and is significantly altered five years after handover, it might be challenging for the new owner to maintain the currency of an historical AIM. In other situations, such as where a building retains its commissioned use over a significant part of its whole life cycle, a federated AIM comprising

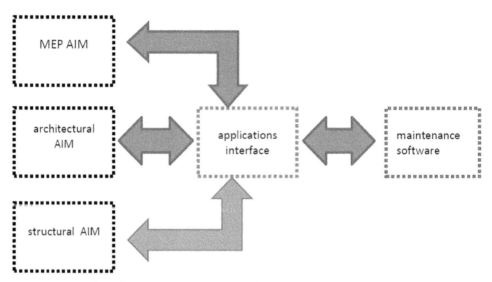

Figure 17.1 Data flows from AIMs to maintenance software

architectural, structural and building services components (Figure 17.1) could be maintained and refreshed over a number of years. In that circumstance (at least theoretically) it should be possible to take a long-range view with functionality of the asset model. In addition to providing a useful resource for servicing planned/reactive maintenance activities, space planning and adaptive reuse, the AIM could also be embedded with data to facilitate safe dismantling and/or recycling at the end of useful service life.

Domain-specific software house Conject Ltd highlighted that the approach taken by some project teams is to use a combination of generic file-sharing systems and email to manage and co-ordinate BIM models and associated data with their project partners. It was argued this methodology fails to meet the requirements of PAS 1192-2 and BS1192:2007 working practices, as digital models are being shared in undefined formats without a 'schedule of information deliverables', adequate change control or audit trails. The company also noted that this 'fractured approach to managing BIM processes and workflows results in ambiguity, inaccuracy and incomplete handover data, undermining the objective of delivering built assets via BIM with structured (COBie format) data drops and/or an Asset Information Model (AIM)' (Anon 2015).

One of the most effective and pre-emptive acts with maintenance operations, BIM and extended time frames is categorising work into classification sets. The fundamental protocol for determining classification types would be the separation of planned and reactive maintenance activities. That action follows a basic and enduring rule of working with digital data. It cross-refers with the structured approach to data management and BIM set out in the BS1192:2007 source document. In practice, the structured rather than arbitrary approach facilitates the quantification of various types of work orders being generated and recording of where/for how long operatives' time is being resourced. Followed up by testing the effectiveness of any operational activities by monitoring work activities performed in each classification. These actions can all be co-ordinated and referenced via a data-driven AIM.

By definition, application of soft landings principles includes a three-year aftercare phase. If the BSRIA five-stage protocol has been embedded in the BIM process during design-construct phases, the final stage (extended aftercare and POE) flags up a number of key points which need to be embedded in maintenance plans and information strategies for implementation.

By the end of Year 1 post occupancy, there is an assumption that any teething difficulties with building performance will have stabilised. For example, in the case of a new-build supermarket, where the heating and ventilation system proved problematic from handover until six months into use, root causes will have been determined and rectified. In Years 2 and 3, the focus should be less with resolving post-handover issues and more on recording operation of the building and reviewing performance in use. By then, the building management/maintenance team (either in-house or outsourced with smaller projects) should be fully in command of managing the building fabric, systems and user needs.

With a BIM-enabled project, key tasks will include reviewing data and operational strategies and, most importantly, ensuring the currency of the AIM as the primary data source for post-occupancy building management. Whether using sophisticated facilities software for large facilities and/or property portfolios, or more basic databases for smaller projects, the fundamental principles underpinning efficient information management are the same: maintaining data in structured and readily accessible formats.

For larger projects, bespoke systems for maintenance management like the Concerto Help Desk are underpinned by rich data and flexible functionality and can be linked to AIMs to offer information management linking customer with support response teams. Centrally managed, the Concerto Help Desk allows users to log calls and track their progress through to resolution. Operatives can manage requests for action via a user-definable web portal. Typically, these actionable instances would include maintenance requests, complaints, enquiries or general comments, along with works ordering and invoicing.

When requests are flagged up, in the interests of transparency, notifications are automatically sent to users. Operatives can then raise work orders, purchase orders, quotes for services and the like. Callers are kept fully updated throughout the process, and the facilities team is alerted to progress of outstanding calls against customer requests. The system uses intuitive, graphical information to offer a quick visual overview of progress and outstanding tasks. Calls can be logged by clients via a web browser on any handheld device, anywhere. Contractors can operate on the move, allowing them to access the information they need and feed back their data from the point servicing without additional administrative actions. An example of a centralised data-driven application being used with a small project would be by utilities providers undertaking planned maintenance of a heating system, where the operative can communicate from the site to the centralised database via a handheld device and a wireless connection.

Islington Council's property department manages a small but very dense area. The local authority migrated from an outsourced system for maintenance work allocation, to contractors. The Concerto Sites data-driven asset management application was already being used by the local authority and, having assessed the system, Islington's Business Continuity Team recommended the software for maintenance management. The ability to integrate the software with smartphones and other handheld devices allowed for greater efficiency and faster response times by contractors. The system can also email feedback to clients and allow users to monitor work in progress.

Ultimately, the value of post-occupancy digital data for building maintenance is primarily about offering efficient connectivity between source of request and destination for action. In that context, the BIM Asset Information Model, however adequately defined and appropriately populated, is only one element in a data-driven process which connects building users with maintenance agencies through enabling systems.

17.5 Planned and reactive maintenance

Although a template for post-occupancy information management in the UK has been established with PAS 1192-3, at the present time the practical implications, both in terms of strategies for planned/reactive maintenance and in effecting related cost/efficiency gains, cannot be articulated with any degree of confidence. As Burcu Akinci and colleagues at Carnegie Mellon University noted (Acamente *et al.* 2010), the current focus for BIM implementation seems to be on the design and construction phases. Utilisation of BIM post occupancy, for example, to facilitate planned and reactive maintenance, seems to be less of a priority for the construction industry. The exploration of how spaces and objects in 3D information models can be enriched with useful data to serve post-occupancy needs is still at an early stage of development.

That emphasis on design-construct stages with BIM practice seems paradoxical when the literature suggests that 60 per cent or more of a building's whole life cost is likely to be expended during the occupancy phase. In that context, it is worthwhile taking a snapshot of some of the issues which come into play after building handover in relation to planned and reactive maintenance. As the processes of transferring information from pre- to post-handover information models mature, the potential for BIMs to support maintenance planning and react to building owner and user needs may be developed more comprehensively through research and practice. Although that strategic direction is already embedded in some organisations, the practicalities of using BIMs for planned and reactive maintenance still requires significant development through concerted dialogue between commissioning clients and their service providers.

Southwest One is a joint venture set up between Somerset County Council, Taunton Deane Borough Council, Avon and Somerset Police and the global IT and business management provider, IBM. Initiated in October 2007, the partnership was projected to run for an initial period of ten years. Key objectives included:

- improving access to and delivery of customer-facing services;
- investing in new world-class technologies to improve productivity;
- updating, improving and reducing the cost of corporate and support services;
- modernising the overall way the partner organisations work.

Planned maintenance is typically work programmed at the beginning of a fiscal period and over a planned cycle of activities, for example, over successive three-year slots. In the case of Southwest One and their property portfolio, programmed maintenance projects are the larger items of work that have the biggest impact on a site, as they have the ability to transform the environment, for example, in the case of a school or similar public building. Typical programmed projects include heating system renewals, external redecorations, electrical rewires, laboratory refurbishments, toilet refurbishments and large-scale window and roofing renewals.

From Southwest One's perspective, there tends to be less urgency associated with programmed projects and therefore more time to pre-plan and ensure that clients get maximum benefit, minimal disruption and value for money from the project. These projects are more complex to deliver than reactive repairs and may require specialist input (for example, 'hard' FM services). Interaction with specialists provides client assurance that the most appropriate solution has been designed; one which complies with statutory requirements and is undertaken safely by a contractor who is being properly managed. Hard FM ensures that the contractors are competent to carry out the work and are appropriately managing the numerous risks inherent in this type of work. These include asbestos, Control of Substances Hazardous to Health (COSHH) and Construction (Design and Management) Regulations.

Southwest One have defined reactive maintenance as the day-to-day work required to correct operational equipment failures and ensure that their properties continue to operate safely and effectively, with minimal disruption, despite unpredicted faults and breakdowns. Reactive maintenance work is often minor, for example, repairing a broken pane of glass, leaking tap or unblocking a drain. From a client perspective, unforeseen maintenance tasks can be difficult to quantify from the outset and expensive to execute and, in some cases, involve serious disruption to continuity of building occupancy. Some situations may also require significant risk assessment prior to implementation at various levels. These could range from a building-specific instance, for example, discovery of asbestos, to area incidents requiring reactive responses to groups of buildings affected by storms, floods and similar climate-related damage.

Figure 17.2 shows a hierarchy of priorities for reactive maintenance items from a university property portfolio structured into four categories. From a client's perspective, in preparing a brief for information requirements for an Asset Information Model (AIM), taking time to develop a structured approach will pay dividends in reactive maintenance operations. With a BIM project, some key information necessary to feed reactive maintenance requirements (for example, COBie data for equipment specifications) may be embedded in the pre-handover Project Information Model (PIM) and will require a transfer process to the AIM. Other data packages, such as method statements for a range of reactive maintenance contingencies and tasks, may be added to the AIM structure after handover. While PAS 1192-3 provides a reference standard, client and organisational needs will take precedence with specific projects.

As a property holder and commissioning client, Kent County Council's Construction, Facilities Management and Assets Teams have a vision that Building Information Modelling (BIM) represents both an enhanced technology and a process change for the authority's

Priority		Nature	Examples
Urgent Priority	1	Life threatening potential	Risk of fire or explosion Gas leaks Interruption of water supply Fire alarm failure Loss of electricity Lift breakdown in use
Urgent Priority	2	Operational difficulties	Blocked drain Heating or ventilation system failure Local loss of power Security system difficulties
Essential	3	Operational difficulties	Partial loss of power in a room Repairs to internal doors Loose fittings/fixtures Replacement sanitary fittings Loose floor coverings not posing H&S risk
Non-essential	4	Do not inhibit operation	Painting defects Sticking doors not posing H&S risk Missing signage Damaged equipment Vandalised fittings

Figure 17.2 Hierarchical typology of reactive maintenance items for AIM structure

property portfolio. The Council is committed to moving both the organisation and its service providers to BIM as quickly, effectively and efficiently as possible. In articulating the fit of BIM within a higher level strategy for sustainable development, the Council already has the infrastructure in place to make that transition from conventional practices for maintenance management. The next step for Kent County Council would be to enable integration of BIM methodologies and process requirements into its delivery requirements for supplying and servicing organisations. As a case in point, it is the transfer of these aspirations into actions and diffusion through supply chains which is likely to have the most significant and pervasive impacts on stakeholders within the authority's supporting networks.

17.6 Embedding the user experience into post-occupancy evaluation

The UK Government's intentions with BIM were set out in the 2011 strategy document and have been followed through by the actions of the BIM Task Group. Embedding Level 2 principles into publicly procured pilot BIM projects, such as the Cookham Wood Young Offenders Institution near Rochester, has provided opportunities for upskilling, bi-directional learning and incorporating lessons learned into evolving BIM processes. In parallel with developing its own activities, the Task Group has made significant efforts to reach out to the industry and promote the benefits of BIM.

To what extent these challenges have been taken up across UK construction is not clear at this point in time. Neither is the impact on the employability of SMEs and micros which make a conscious decision not to engage with BIM processes, techniques and workflows. As some regional and local authorities have already followed the Government's lead in setting up protocols for BIM-enabled publicly funded projects, the consequences of processes initiated from the centre could have wide-ranging impacts on small organisations over the next few years.

Embedding the building user perspective into the post-handover phase is a tried-and-tested facet of post-occupancy evaluation (POE) as a technique to gather reliable and useful data about building performance in use. If a soft landings philosophy is applied as a mandatory project outcome, a POE marks a milestone in the process of user input from early design stages forward. With BIM scenarios there may be checks incorporated into the design development process to monitor user interactions. These checkpoints can be mapped and audited against the EIR, BEP and other process management templates as a project develops. For example, in the situation where the RIBA Plan of Work 2013 is being used as a project model, a sustainability gateway can be embedded at the end of each work stage both to test/validate the incorporation of 'hard' (e.g. predictive energy performance) and 'soft' (e.g. user feedback) criteria into the design development process, and to identify data flowing into and out from the BIM environment as required to populate predictive analysis techniques such as energy performance appraisal.

Between 2010 and 2011, the Department of Education for England's Partnerships for Schools (PfS) initiative carried out a POE of 25 schools in England. The sites included 9 primary schools, 14 secondary/sixth-form schools and 2 special schools, all of which had benefitted from a new build or major refurbishment project (over £500,000). This evaluation was a development from a pilot POE project completed in 2009–2010 and combined feedback from peer professionals, staff and students with an assessment of energy performance. A number of performance criteria were flagged up for testing. These included peer and user perceptions of quality of built environment, internal space organisation, safety/security and long-life/loose-fit qualities. The key strategic objective was to create schools which could adapt and evolve in the future.

Transposed into a BIM environment, the methodology for a similar but more direct POE exercise could be set up with building users recording/sending data using interactive techniques (for example, a touchscreen/graphic display input source) to facilitate dialogue and information logging between end user and an Asset Information Model structure. As Mark Klimt of law firm DWF noted in the context of occupant critique of schools in use, 'the students will be robust and unforgiving end-users whose activities will demand a lot from their environment' (Klimt 2015). Once the information had been logged, integration with the building management process and translation into decision-making and appropriate actions (be it space reorganisation, changes to equipment logistics, or environmental factors) become a straightforward process which can be initiated using the AIM as a resource for facilitating performance-enhancing improvement.

Ibrahim Motawa and Wendy Corrigan (2012) argued that despite growing interest in BIM, BIM-driven post-occupancy evaluation of buildings is not an established technique at this point in time. As has been noted elsewhere, practical applications of BIM are mainly visible at design and construction stages. There is limited research linking knowledge gained from energy practices with BIM models to improve the user responsiveness and energy efficiency of buildings. One exemplar from the literature was the Scottish Government-supported CIC Start Online programme, which aimed to embed sustainable building design into practice to assist SMEs and micros to develop, prototype and test innovative ideas. Motawa and Corrigan's 2012 project on post-occupancy evaluation also noted that utilisation of BIM for POE could enhance the capability of digital tools to enable an holistic and fully integrated process, in particular in supporting stakeholder collaboration by offering the facility to insert, extract, update and modify facilities information held in BIM models.

The 2013 Cabinet Office report on soft landings identified that a key aspect of the BIM Task Group's legacy phase would be to ensure that the Government Soft Landings (GSL) process was applied to all central government projects through briefing and design development taking account of end user requirements. The aspiration was to put in place protocols to enable publicly procured projects became more operationally efficient, effective and responsive to user needs. Enabling that process demands a sharp focus on performance requirements from the outset. In that context, the shape of GSL as a variant of the long-established BSRIA 'soft landings' protocol has already been discussed.

The GSL guidance documents have been structured to provide appropriate information to ensure that performance requirements are translated into specific targets which are assessed regularly during project evolution and post completion. The functionality and effectiveness of predicting attainment of these performance targets can be tested by quantitative and qualitative data gathering and analytical techniques using appropriate POE techniques. BSRIA and the Usable Buildings Trust have a significant track record of activity in that field.

Among published soft landings case studies, the design development phase of services engineers Max Fordham and Partners' input into the Keynsham Town Hall new-build project incorporated workshops being held with clients and with technical managers. The purpose of the dialogue was to gather information on soft landings requirements which were incorporated into the contract documents. That instance marks a significant difference in practice between 'guidance for information' and a tougher performance-based strategy where implementation is embedded into formal agreements between client and design-construct team members.

For BIM projects configured to adopt the PAS 1192-2:2013 protocol, the content and fit of post-occupancy data within an AIM should reflect the client's information requirements as determined at briefing stage. In terms of interactions with developing and implementing a soft landings strategy, the frequency and content of information exchanges (data drops) during design evolution will also be identified at project inception. Key functional/user-specific needs, information requirements and methods for verification will be written into project documentation (EIR and BEP). In the situation where the RIBA Plan of Work 2013 template has been adopted, procedures for verifying that the soft landings process is being followed can be embedded at the conclusion of each work stage, cross-referenced with BIM documentation and carried forward to the post-occupancy phase to provide benchmarking data for review by POE from within an AIM.

Criteria to test the performance of a project after handover may include an assessment of how well the building responds to the users and facilities management team's requirements for functional outcomes, such as:

- human comfort factors: temperature, air quality, lighting, noise and control;
- functional performance of individual spaces;
- amenity of spaces: access and egress, quality of horizontal and vertical circulation, interactions between spaces;
- user impressions of spaces: psychological factors such as perceived quality of rooms;
- maintenance: access and ease of interaction with equipment for servicing and replacement;
- durability of equipment; and
- health and safety of building operators and users.

At strategic briefing stage, it is important that post-occupancy performance criteria are discussed with the commissioning client and translated into a specific set of information requirements. As has been discussed, that may be just one of a number of data sets incorporated into a project EIR.

It has already been highlighted, in relation to the management of information streams generally, that one of the complexities with applying BIM to projects is the need to reference a number of industry paradigms for managing process. Typically these may include PAS 1192-2:2013, BS1192:2007, RIBA Plan of Work 2013, NBS BIM Toolkit, NRM 1,2,3, BSRIA soft landings templates, and others. For an SME or micro organisation that may represent a challenging mix on top of other regulatory standards such as Construction (Design and Management) Regulations 2015 already in the frame.

Central government published sources on BIM and digital process do exhibit a tendency to focus on information management to the exclusion of other aspects of the complex procedural and people interactions embedded in the processes of making buildings. Ultimately, soft landings and involving users in the pre-occupation stages of procurement processes is about producing better and more user friendly buildings. Ironically, the Cabinet Office strategy for the implementation of GSL does seem to reflect that aspiration.

At project level, it is essential that templates for doing BIM are applied holistically. However rigorously robust and efficient information management is incorporated into process, that in itself will not produce better and more responsive built environments.

Bibliography

Acamente, A., Akanici, B. and Garrett Jnr, J.H. (2010) 'Potential Utilisation of Building Information Models for Planning Maintenance Activities', *Proceedings of the 2010 International Conference on Computing in Building and Civil Engineering*, Nottingham University Press, [online]. www.engineering.nottingham.ac.uk/icccbe/proceedings/pdf/pf76.pdf [accessed 18 March 2015].

Anon (2011a) *BIM Management for Value, Cost and Carbon Improvement*, A report for the Government Client Construction Group, Department of Business Innovation and Skills (BIS), [online]. www.bimtaskgroup.org/wp-content/uploads/2012/03/BIS-BIM-strategy-Report.pdf [accessed 11 February 2015].

Anon (2011b) *GSA BIM Guide for Facility Management*, General Services Administration, [online]. www.gsa.gov/portal/content/122555 [accessed 11 February 2015].

Anon (2012) 'Post Occupancy Evaluation of Schools 2010–11', *Building*, 26 April, [online]. www.building.co.uk/Journals/2012/04/26/o/q/p/POE-full_report.pdf [accessed 12 March 2015].

Anon (2013a) *Government Soft Landings; Section 2 GSL Lead and GSL Champion*, Cabinet Office, [online]. www.bimtaskgroup.org/wp-content/uploads/2013/05/Government-Soft-Landings-Section-2-GSL-Lead-GSL-Champion.pdf [accessed 12 March 2015].

Anon (2013b) 'Meadow Road Housing Project, Worthing', *BIM fusion*, The Clarkson Alliance, [online]. www.bimfusion.co.uk/contact/ [accessed 12 March 2015].

Anon (2014) *Government's Estate Strategy*, Cabinet Office, HM Government, [online] https://www.gov.uk/government/uploads/system/uploads/attachment_data/file/360262/Government_estate_strategy.pdf [accessed 11 February 2015].

Anon (2015) 'Collaborative Building Information Modelling BIM', Conject, [online]. www.conject.com/uk/en/use_cases-bim [accessed 19 March 2015].

Anon (n.d.a) 'Case Study Facilities Management', Concerto, [online]. www.concerto.co.uk/wp-content/uploads/2014/01/FACILITIES.MANAGEMENT.ISLINGTON.pdf [accessed 19 March 2015].

Anon (n.d.b) 'Planned and Reactive Maintenance', Southwest One, [online]. www.southwestone.co.uk/welcome/shared-services/property-services/projects-and-technical-resources/design-services/facilities-management/planned-reactive-maintenance/ [accessed 19 March 2015].

Beighton, A. (2013) 'Changing Processes in Reality with BIM', *Construction Manager*, 23 August, [online]. www.construction-manager.co.uk/management/how-bim-will-change-processes-and-behaviours-reali/ [accessed 12 March 2015].

Dawes, L. (2014) 'BIM: Reducing Complexity without Losing Clarity', *CIC Blog: BIM Research*, CIC, 23 January, [online]. http://cic.org.uk/blog/archive.php?cat=bim-research [accessed 12 March 2015].

Freeman, M. (2013) 'Gut Feel or Hard Numbers? The Importance of FM Data', *The Mitie Debates*, Mitie, 8 October, [online]. www.mitie.com/blog/the-mitie-debates/gut-feel-or-hard-numbers-the-importance-of-fm-data [accessed 11 February 2015].

Gough, T. (2014) 'Building Information Modelling and Kent County Council', Kent County Council, [online]. www.secbe.org.uk/documents/terry_gough_kcc_presentation_2014_compatibility_mode.pdf [accessed 19 March 2015].

Jackson, R. (2014) 'openBIM for Facilities Management', *BIM Blog*, Bond Bryan Architects, 7 November, [online]. http://bimblog.bondbryan.com/ [accessed 11 March 2015].

Klimt, M. (2015) 'Legalese', *The Architects Journal*, 241(10), 13 March.

Lake, A. (2013) *The Way We Work: A Guide to Smart Working in Government*, HM Government, [online]. www.niassembly.gov.uk/globalassets/documents/finance/inquiries/flexible-working/research-papers/cabinet-paper-on-a-guide-to-smart-working-government.pdf [accessed 11 February 2015].

Lewis, A., Carrasquillo-Mangual, M., Cocherl, T. and Bogen, C. (2013) 'March 2013 COBie Challenge for Facility Management', *Proceedings of National Facility Management & Technology Conference*, buildingSMARTalliance, [online]. www.nibs.org/?page=bsa_ccfms13 [accessed 12 March 2015].

Motawa, I. and Corrigan, W. (2012) 'Sustainable BIM-driven Post-occupancy Evaluation for Buildings', *CIC Start Online*, [online]. www.cicstart.org/userfiles/file/FS-49-REPORT.PDF [accessed 12 March 2015].

Way, M. and Bordass, W. (2014) *The Soft Landings Framework*, Usable Buildings Trust and BSRIA, [online]. http://usablebuildings.co.uk/UBTOverflow/SoftLandingsFramework.pdf [accessed 11 February 2015].

18 Feedback and feed forward

18.1 Review of project performance and feedback into organisational structures

Patrick MacLeamy of HOK argued that for clients and building owners, the financial out-goings required to run and maintain a building over its whole life period can be five times the cost of design and construction combined, suggesting a corollary that efficient use of data over project life cycles offers the potential to save a substantial part of future design and construction costs. Clearly the BOOM (building operation optimisation model) phase from MacLeamy's BIM, BAM, BOOM whole life paradigm could have a significant impact on the attainment (or not) of a range of post-occupancy factors (heating, services, building maintenance costs, and the like). Extrapolating the premise from a single building across an industry suggests that while attainment levels with UK Government 2025 targets may not be quantifiable at this stage, the post-occupancy phase may offer the greatest potential for realis-ing performance gains with BIM. Yet in practice there may be a disconnect between pre- and post-handover phases of a building's whole life evolution and management.

The Probe (Post Occupancy Review of Building Engineering) series of investigative stud-ies demonstrated convincingly that some buildings do not perform in use as envisaged by design teams. Perhaps, in some cases, performance criteria for the post-occupancy phase of a building's whole life evolution were not formalised and embodied into the design brief. Typically, inadequate performance in use could impact on running costs, client/user satisfac-tion with facilities, planned/reactive maintenance overheads, and the like. For clients such as housing associations, managing both existing stock and rolling programmes of new build and maintenance, effective feedback from shortcomings in design, commissioning processes and buildings in use can improve satisfaction levels among users and enhance the overall quality of service provided. That feedback needs to be incorporated into front-end briefing strategies for design and data management.

Kingdom Housing Association's vision is to be a leading provider of quality affordable hous-ing in Scotland and to be recognised as offering innovative and excellent service. Kingdom lists innovation and commitment to teamworking among its core values. In 2012, the housing association undertook a 27-unit innovation showcase development on a greenfield site out-side Dunfermline for affordable social rented housing. A project partnering arrangement was used in an effort to achieve supply chain integration between suppliers and a range of con-tractors developing house types over the ten sites comprising the demonstration project. The client identified potential common product suppliers who could be used by all developers to standardise components such as windows, doors, kitchen units and roof tiles.

A fundamental driver was to incorporate POE into the post-occupancy housing management process. The monitoring and evaluation programme was divided into two stages: Stage 1 'post construction/early occupancy' and Stage 2 'in use and post occupancy'. Data sets from the Stage 1 collection phase included:

- field study results which evaluated the as-built thermal performance of walls, ceilings and floors;
- a design and construction audit to all dwellings, including a comparison of the design outputs against the as-built outputs;
- a full review of SAP worksheets;
- a system performance evaluation of low-carbon technologies;
- an infrared thermography of all blocks;
- an air tightness evaluation of all blocks;
- an acoustic performance evaluation of all blocks;
- an as-designed/as-built comparison of predicted against actual energy usage; and
- a survey of resident satisfaction levels with various aspects of design, including comfort levels.

A key finding impacting on the organisation's feedback/feed forward protocols was that building performance energy simulation (BPES) tools should be embedded in the design process. Also, perhaps more significantly for design teams, greater effort was needed to try to close the gap between predicted and actual energy performance in use. In that context, Baba and others (2013) have advocated a staged approach to prototyping, appraising and validating energy appraisal data during outline and detail design phases. Using a BIM paradigm for reference would suggest a series of predictive energy iterations as a design developed over time. To highlight that point, Kingdom's POE found that in one case, energy consumption for space heating in use was close to double the 'as-designed' predicted values.

The Kingdom exemplar again begs significant questions regarding the aggregation, accuracy and bi-directional capabilities of data to serve post-occupancy requirements. Lachmi Khemlani (2011) cited the EcoDomus FM software as a platform which could integrate various data strands developed over the whole life of a project, including GIS systems, BIM models and FM systems. Reference was also made to building automation systems which could facilitate operational data capture and analysis to facilitate providers like Kingdom being able to take a more holistic information-centric approach to performance management of their housing stock. In linking client information needs with PAS 1192-2-driven BIM scenarios, a method for identifying, developing and managing these data sets needs to be embedded at the front end of projects. Early client/design team conversations form an essential part of the briefing process and are central to ensuring that critical information requirements are identified early on, linked with the efficient management of digital data and targeted towards the achievement of required post-occupancy outcomes.

18.2 Using and maintaining the currency of the BIM inventory, PAS 1192-3

As Lachmi Khemlani also noted, organisational capability for BIM integration does not mean that it is actually being implemented, particularly during the post-occupancy phase of a project's life cycle. For example, in the case of a small design firm using lonely BIM for pre-construction documentation, information locked into a data-rich BIM model (and/or a suite of project models) represents a potentially valuable resource for a client organisation well beyond requirements to service the build process. In practice, the potential which that resource offers may be significantly under utilised unless the designer's BIM capability is embedded into more inclusive methodologies to use BIM for collaborative working across the broad scope of design/construct/maintain processes. That situation flags up a couple of key questions.

First, how can information embedded in a data-rich model during design and construction be captured and migrate to a post-occupancy data inventory to service planned, reactive maintenance and other client requirements? Any one of a number of methods could be adopted to set up a post-occupancy inventory. These could range from a manually updatable spreadsheet, through use of data-rich 3D parametric models, to the adoption of sophisticated facilities management software tools. As with all things BIM, there is no single way. The UK Government's preferred COBie protocol for data management may in theory be viable, but in practice is too unwieldy for some small organisations to implement.

Second, digital data sets which migrate and morph between design and post-occupancy phases need to be accurate, fit for purpose and up to date. That triangulation of data raises huge challenges for collaborative teamworking over time, particularly in relation to how the data is going to be managed and by whom. As was highlighted in Section 16, a wall-hung domestic boiler is an example of an object which could start its digital journey as a geometrical representation within a 3D model at early design stage and reach post occupancy as a manufacturer's BIM object carrying embedded instructions (metadata) for commissioning, user instructions and maintenance. Once the actual boiler has been commissioned, service technicians are able to run in situ testing digitally linked to manufacturer and service organisation databases. During the early stages of post occupancy, these requirements pull together clients, design-construct teams and building users into a single frame of reference. While it is unlikely that a client for a single family house would be investing in FM software, the principle of assimilating and managing appropriate data to serve post-occupancy needs applies universally across all building types and sizes. Building users need to be involved in BIM processes which may impact on post-occupancy outcomes. For example, as Barry Connolly of real estate consultancy JLL claimed, there could be a significant disconnect between a building's developer and its end occupier. Connolly reasoned that if a tenant knew a developer could save 30 per cent on energy costs by specifying certain equipment or materials, the combination of that knowledge and the tenant's engagement during the design phase could influence the decision-making process and practical consequences.

As a protocol to frame the use of digital data during post-occupancy building management, PAS 1192-3 sits within the PAS 1192 suite of reference documents. PAS 1192-3 cross-refers with the BS ISO 55000 series of standards and the earlier PAS 55-1 and PAS 55-2. The intention was that these documents would provide one overarching framework for the adoption and implementation of PAS 1192-2 and PAS 1192-3. As with many aspects of the PAS 1192 suite, it can be quite daunting for a small organisation to absorb the full scope of the documentation, then decide how PAS 1192-3 could usefully be applied in practice. The process of translation, filtering and application will invariably be time-consuming first time round. BIM evangelist Kath Fontana, managing director of BAM FM, argued that because periodically FM contracts turn over and operational data held by the outgoing contractor is returned to the client, there is a significant risk in relation to the accuracy of the data set. An incoming contractor might then want to re-evaluate that data, invariably leading to a higher contract price. The existence of an operational building information model with embedded data would reduce exposure to that risk.

The essence of PAS 1192-3 is establishing a framework for data management during the post-occupancy phase of an asset's life cycle. The term 'asset' is used as an umbrella term to cover references to buildings and infrastructure. In the context of a BIM whole life scenario, PAS 1192-3 sets out typical data transfer processes to:

- create an Asset Information Model (AIM);
- migrate data from a Project Information Model (PIM) to the AIM;
- use and maintain the currency of the AIM as a change management tool;
- record information relating to the disposal, decommissioning or demolition of an asset; and
- hold the AIM as an organisational resource.

Paradoxically, PAS 1192-3 distinguishes between 'asset management' and 'facilities management'. For a small organisation, that distinction may seem superfluous, particularly when the 'hard' (building fabric) and 'soft' aspects (space/people management, catering, cleaning, etc.) of post-occupancy building management may be wrapped up and directed from within a single organisational portfolio. In taking an holistic view of post occupancy, PAS 1192-3 seems to fall short in key areas of building management, and that aspect might limit its usefulness in the field. All buildings change over time; organisational and people needs can be significant drivers for change. As the writer and polemicist Stuart Brand noted 'commercial buildings have to adapt quickly, often radically because of intense competitive pressures to perform … commercial buildings are forever metamorphic' (Brand 1994: 7).

A small organisation bidding for a share of a Level 2 BIM project might reasonably be expected to demonstrate knowledge of the PAS 1192 suite of standards as part of a prequalification process. PAS 1192-3 does, however, beg some questions regarding its inclusiveness. That caveat needs to be borne in mind when developing strategies and operational practice for post-occupancy data management.

18.3 Capitalising on BIM as a medium for cross-disciplinary partnering

CIC Chief Executive Graham Watts described BIM as 'a real force for collaboration because it can't really operate unless you have the entire team on board at the earliest possible stage which encourages much earlier contractor involvement' (Scott 2014). If then, as the UK Government BIM Task Group claims, BIM is 80 per cent people/process and 20 per cent technology, applying the hypothesis suggests that collaboration is the key ingredient (possibly the pivotal element) in realising the UK Government's own formula for digitising UK construction. In fact, if a coalition analogy was to be used, collaboration appears to hold the balance of power as a change agent. Yet ironically, so much of the contemporary BIM discourse focuses on technological aspects of the genre: data-rich BIM authoring packages, digital design/project/document management tools, analytical software, file transfer between BIM software platforms, and the like. Despite the existence of robust standards such as those developed by the buildingSMART® alliance, that diversity in itself may introduce numerous inherent barriers to digital collaboration for SMEs and micros. Small organisations need to be able to develop BIM projects and manage their data flow with a small palette of reliable and affordable digital tools.

In making Level 2 BIM viable for SMEs and micro organisations and to broaden the scope of uptake, the spotlight with education and practice information providers needs to shift from interpretation/application of standards and technical solutions/fixes towards softer organisational and people-related aspects of collaboration. That is a fundamental premise which requires to be addressed by UK industry, professional bodies and educators before significant progress can be made. Californian architect Jonathan Cohen outlined a two-tier hierarchy for project teams to consider in applying the principles set out in the AIA's integrated project delivery (IDP) approach. First tier, core principles were:

- early involvement of key participants;
- shared risk and reward;
- multi-part contract;
- collaborative decision-making and control;
- liability waivers among key players; and
- jointly developed and validated project goals.

Second tier, highly desirable characteristics were:

- mutual respect and trust across the team;
- collaborative innovation (new ways of achieving familiar outcomes);
- intensified early planning;
- open communication within the project team;
- applying lean principles to design/construction/operation; and
- interdisciplinary troubleshooting (application of Toyota's 'Big Room' principle);
- transparent finances.

(Cohen 2010)

Filtering out the contractual and liability aspects from the core principles and substituting 'mutual respect and trust' may offer an appropriate template for cross-team collaboration, in effect serving as an entry-level protocol for small organisations to apply IDP principles to projects developed within a BIM environment.

Considering another aspect of partnering, if the perspective shifts from 'where are we now?' to 'where do we need to be five years ahead?', the direction of travel with undergraduate education needs to be in the mix. SME and micros recruitment requirements will be for graduates who are 'BIM ready', not only being able to demonstrate discipline-specific attributes, but also having underpinning knowledge of BIM in the context of pan-discipline teamworking. Because of the way in which built environment undergraduate higher education tends to be structured and contained within discipline silos, embracing pedagogies for multidisciplinarity with BIM could be a tall order for UK industry to deliver on. Plus, in terms of people resources, industry commentator Su Butcher, quoting from the CITB's annual Construction Skills Network report, tweeted in January 2015 that an additional 44k people are required every year in construction just to keep up demand.

At a CIC workshop in 2013, Richard Saxon, UK Government BIM Ambassador for Growth, argued that BIM can't be taught with no read across disciplines and suggested, for example, 'crash weekends' where students from different disciplines work together solidly on the same multidisciplinary project. He argued that reorganisations in higher education institutions could position construction and built environment undergraduates away from each other and could 'deepen' silos, which would not be helpful in learning how to apply BIM across discipline boundaries.

Certainly, there have been some initiatives. Since 2011, the UK BIM Academic Forum has evolved as a forum to encourage interaction between academic institutions. In Australia, Jennifer Macdonald has developed the IMAC framework as a tool to encourage educators to benchmark their curricula and develop strategies for embedding BIM into undergraduate education. Meanwhile, the UK Government's priority in addressing 2025 targets has been to develop E4BE as a tripartite and strategic grouping representing the interests of professional bodies, higher education institutions and employers. While BIM is a key theme, the overarching priority is to shape capability and build capacity in the sector. With the number of undergraduates enrolling on construction courses having nearly halved between 2008 and 2014, that high-level objective may prove difficult to realise unless the trend reverses. Many built environment SMEs and micros already have close links with universities through tutoring and professional body interactions. In the short term, these informal partnerships and networks may prove more effective catalysts for change. Time will tell.

18.4 Maintaining a competitive edge

As Diana Limburg has pointed out, investing in digital technologies can represent a significant business overhead for SMEs. Purchasing computer hardware and software can be risky and anticipating the right moment to upgrade existing equipment can be a difficult call to make. For example, operating systems migrating from one release to the next can flag up challenging questions for existing software and hardware installations. Backwards compatibility can be a particular niggle with software. As with most things technology driven, there is no steady state as change is continuous and can be unpredictable. Windows and Apple operating systems tend to have their respective aficionados, sometimes making cross-platform dialogue tricky. Small organisations running stand-alone PCs, laptops or local networks might not have the resources (or inclination) to employ IT specialists in-house and may have to be largely self-reliant for troubleshooting. Typically, SMEs and micros may rely on a key staff member who has developed broad-brush computing expertise. That person's departure can result in a frustrating knowledge and resource gap for an organisation.

Technological change tends to be rapid and unpredictable, and obsolescence can creep up without much warning. Second-guessing next-generation developments of computer hardware and software can be difficult for practitioners and small business organisations. 'Will we upgrade now or wait for another 6 months?' may be a common and germane question for a small organisation to ask. The laws of serendipity may trigger unplanned but fortuitous upgrades, for example when a PC or laptop fails with little or no warning. Thinking more strategically, the prospect of an organisation's first BIM project being won and going live might help accelerate the decision-making process. In that situation, organisational levels of knowledge, understanding and experience with BIM are likely to be tested, particularly if a prequalification process is a gateway to winning work. For an organisation to profess BIM knowledge and expertise, without the confidence of being able to deliver on the substance, is a high-risk strategy.

In the last few years, there has been a gradual shift from software being purchased and resident on office-based hard drives towards the use of cloud-hosted subscription services. For those who might believe that 'the cloud' is a moniker for next generation technology, in fact it is just another metaphor for the Internet. The transitional trend towards remotely hosted software can present a mixed bag of pros and cons for a small business. As a case in point, Limburg (2012) argued that outsourcing could dull any competitive edge gained from information technologies and lead to overdependence on suppliers. For small businesses with an inclination to rely on the cloud, Susan Ward flagged up five caveats:

- Possible downtime because cloud computing increases dependency on the Internet. When it is offline, or when viability of a wireless is intermittent, it can be disruptive if the business relies on cloud-hosted software to function.
- Cloud computing also means that data is stored remotely and to some extent controlled by the provider. Reports in 2013 that the US National Security Agency were able to tap into data circulating between user accounts and the world's largest tech companies were not reassuring. When one of Adobe's servers was compromised in 2013, it was reported that encrypted data of over 2 million customers could have been vulnerable to access by unauthorised users.
- Monthly or annual subscription costs of cloud-based subscription services need to be balanced against one-off software purchases. While subscription costs may be higher per annum, the currency of the software is less of an issue than when purchased outright.
- Care needs to be taken when signing up with cloud-based providers that the user is not being locked into proprietary applications or formats.
- Some cloud-based applications may not be particularly responsive in providing customer support. Not every business is prepared to tolerate non-productive time allocated to trawling through user forums to resolve software issues.

(Ward 2015)

But there may also be counter arguments in favour of cloud computing as a resource for small organisations. For example, as Limburg noted, by using web-hosted services, SMEs can rent the latest and most appropriate applications. These can be scaled up as required, without capital investment in software which could become obsolete within a few years, allowing SMEs to make better use of IT for competitive advantage and supporting the idea that engagement with web-based technologies provides a significant assist in allowing SMEs and micros to punch well above their weight with BIM. Limburg also cited Webb and Schlemmer in arguing that while information technology is relatively cheap, ubiquitous and accessible, competitive advantage is not gained by simply having it, but necessarily has to derive from organisational capabilities to effectively exploit and manage the potential which digital technologies can offer to add value to business development.

London architect David Miller, writing in 2012, clearly held the view that for his SME practice which grew from 4 to 14 staff over three years, the payback from BIM outweighed any inherent challenges, particularly in being able to develop a competitive edge in a difficult financial market. For David Miller Architects (DMA), possibly the most significant overheads in migrating to BIM were capital expenditure on equipment and staff training costs:

A realistic budget for a workstation is around £10,000 once you include hardware, software and training. Even when offset against the cost for a conventional workstation it represents a big investment. DMA have spent £30,000 a year since 2007. However viewed in relation to technical staff costs and fee income, it starts to look a little less scary. We believe that if you have high quality professional staff it makes commercial sense for them to be using the best available tools.

(Miller 2012)

As discussed earlier, DMA reported that key benefits from engagement with BIM included greater consistency in output (which was generating repeat business), greater clarity from client consultations (particularly during the early stages of design development) and simplified workflows for internal design management. However, in terms of forward planning, it is not always possible to measure gains in advance of changes to work practices. In that situation, a process of incremental change (as DMA adopted) may help to mitigate risk.

18.5 Learning and informing future project development

The Organisation for Economic Co-operation and Development (OECD) is a forum where the governments of 34 democracies with free-market economies work together with other countries to promote economic growth, prosperity and sustainable development. In contemporary built environment discourse, BIM seems to be manifest as an industry-specific subset of the commonly cited 'knowledge economy' which the OECD identified in 1996 as a consequence of fuller recognition of the role of knowledge and technology in economic growth. Knowledge, as embodied in human beings (described by OECD as 'human capital') and in technology, has always been central to economic development. But only over the last few years has its relative importance been recognised, as that significance continues to grow. The OECD economies are more strongly dependent on the production, distribution and use of knowledge than ever before. Output and employment tends to expand fastest in high-technology industries, such as the computer, electronics and aerospace industries.

For built environment players somewhat further down the technology chain than aerospace, holistic engagement with digital technologies (the term 'BIM' is a convenient if slightly blurred catch-all) is pivotal to development and, ultimately, survival in today's knowledge-based economy. That hypothesis can be applied equally to macro- and micro-organisational structures. At the SME and micro end of the construction industry spectrum (in terms of organisational size), the argument is a central driver for raising awareness of the need to review business models, innovate as appropriate and engage with BIM. The writer Chris Anderson condensed his take on the diffusion/absorption of two decades of digital technology innovation into two sentences: 'The past ten years have been about new ways to create, invent and work together on the Web. The next ten years will be about applying these lessons to the real world' (Anderson 2012: 36–37). While as an industry, we can debate and chew over various interpretations and consequences of BIM, clearly the phenomenon is not going to go away.

Economic geographer Gernot Grabher projected the view of knowledge as the most powerful engine of economic progress and competitive advantage of our age. 'Knowledge it seems has become magic ... a polyphony of different disciplines has grown to reiterate that our economy has shifted from primary and secondary production to an increasingly knowledge-intense service economy' (Grabher 2004). Grabher also highlighted the shift from the traditional science-driven institutional framework to knowledge production in the context of its application. Each particular context of knowledge application implies a particular set of theories, analytical strategies and learning practices which may not be comfortably transposed onto established disciplinary maps. Having said that, the firm still epitomises the basic analytical building block of the knowledge-based economy. Small firms coalesce, form communities and, as was argued in a European context, should represent a powerful socio-economic force for change.

As construction SMEs and micros look forward from the last project to the next, reviewing project feedback, business models and seeking operational efficiencies may be key (and often underutilised) threads of enquiry. In that context, Stuart Macdonald and others (2000) presented a scenario tagged the 'information technology productivity paradox' as a paradigm questioning the relationship between IT investment and performance, between input and output. Macdonald noted that public discussion of the issue could be dated, with some precision, to a book review by Robert Solow published in the *New York Times* in July 1987. Solow's critique included his eponymous line, 'we see the computer age everywhere except in the productivity statistics'. Perhaps Solow's aphorism line suggests a prompt for SMEs and micro businesses to review the status quo against potential capabilities with BIM. Informed decision-making needs to be triggered from within organisations driven by current business needs and future aspirations.

Within the UK construction scene, the question of how BIM is being measured across the industry has been raised by Peter Trebilcock of Balfour Beatty and others. For example, Tim Platts, Chair of the BIM4SME group, pointed out that much of the construction supply chain was awaiting better evidence for the investment case. 'There are some metrics – for instance there can be a 90% reduction in requests for information – but there is no standardisation about how we measure BIM in UK. This is especially important for SMEs' (Knutt 2015).

In the broad context of information and communication technologies and a boundary condition defined by the organisational map of UK construction (characterised by a plethora of small firms), industry strategies for dealing with the transdisciplinary aspects of BIM are clearly in their infancy. The contemporary focus on standards and protocols possibly disguises more significant but latent challenges of roles, discipline identities and interrelationships which may require deeper enquiry and resolution over the next ten years. Understandably, the priority of small firms may be to focus on the portfolio in hand: the last, current and future projects. BIM may or may not currently fit within that frame of reference. But, referring back to Caroline Stockman's trilogy of aphorisms for remaining competitive in tough times, Lesson 3 (be wary of the status quo) applies without qualification.

If the firm is analogous to the DNA (the fundamental building block) of UK construction's organisational ecology, within firms are project teams and projects, both of which are transient variables over time. The key driver for project teams is to service the requirements of clients and commissioning organisations. As Grabher noted, people come together in temporary work teams and networks which may disappear when a project has been realised. Projects develop around a core team; each team member may bring a different skill set to the table for the project. These skill sets are complementary, and that is the essence of team performance. Because BIM is not yet a ubiquitous activity for UK construction teams, one immediate challenge with collaborative BIM is that firms may come together in groups where BIM knowledge and experience levels are different. That implied lack of equilibrium suggests that BIM will be work in progress for the foreseeable future in the UK.

As Grabher also pointed out, transdisciplinary heterogeneity is a transient characteristic for projects, project-dedicated teams and firms. Cross-disciplinary knowledge generated through project level collaboration may dissipate once a project is completed and design/construction teams are demobilised. In terms of moving BIM documented projects from design/construct to post-occupancy phases, different teams may be responsible for the aggregation and management of data sets. On the other hand, some cross-disciplinary alliances between SMEs and micro-firms may be robust and sustainable over time. That sustainability may be underpinned by mutual trust and respect to a level which transcends the actual methodologies used to develop and deliver projects. Supra-firm alliances may include same or cross-disciplinary networks. Discipline-specific coalitions will invariably include educational and professional body organisations, all with mindsets and strategies relating to how BIM pedagogies and practice should be developed and diffused across their networks.

UK construction is a complex mix of players, organisational sizes and discipline interests. The identity of the small firm may be as a nanosized cog in a giant galactic wheel. On one level, that could be a reasonable analogy. Viewed in the round, the industry fragmentation suggested by Latham and others may not be an inherent barrier to more universal engagement with BIM among SMEs and micros. To paraphrase the neuroscientist Susan Greenfield, if enduring networks of connections between firms could drive a more extensive coalition between organisations, realisation of that process might lead to greater connectivity across the construction industry. In that scenario, it is the small firms, not the 'big beasts' (as Jack Pringle of CIC described large construction companies), which are necessarily the prime movers in stimulating more comprehensive and diffuse industry change with BIM.

18.6 Postscript

The UK Government's 2025 vision for construction flagged up four ultra-ambitious targets for the UK industry:

- a 33 per cent reduction in both initial expenditure and whole life cost;
- a 50 per cent reduction in the overall time elapsed between inception and completion for new-build and refurbishment projects;
- a 50 per cent reduction in built environment greenhouse gas emissions; and
- a 50 per cent reduction in the trade gap between imports and exports of construction products and materials.

In 2011, the UK Government Construction Strategy mandated the use of Level 2 BIM on all public sector projects by 2016. That decision resulted in a Government push to upskill the construction industry with the intention of reducing the capital and revenue costs associated with the procurement and use of buildings and infrastructure. To what extent that initiative has translated into practical actions across the industry is difficult to articulate, particularly among the 250,000 or so SMEs and micro-SMEs forming the backbone of UK construction.

Certainly from the Government side (Cabinet Office), BIM was identified as a significant contributor to the £804 million savings in 2013/2014 construction costs. The Ministry of Justice identified BIM as having enabled £800,000 of savings in the development of the Cookham Wood Young Offenders Institution. And the Government view was that BIM was central to the development of major new infrastructure projects like Crossrail and HS2.

The Digital Built Britain strategy, launched by the Government in February 2015, took the next step towards integrating BIM technologies and transforming the UK's approaches to construction and infrastructure development and construction. The principal aim was to make digitised design and construction processes the norm, and ramp up innovation so that the UK's expertise in cutting-edge digital technologies could capture a significant share of the $15 trillion global construction market forecast by 2025.

The Government was resolved to bolster support for digital technologies in construction by putting in place a number of key measures, including:

- creating a set of new, international standards for open data which would facilitate data sharing across the construction market;
- establishing a new contractual framework for projects which have been procured with BIM to ensure consistency, avoid confusion and encourage open, collaborative working;

- creating a cultural environment which is co-operative and seeks to learn and share;
- training public sector clients in applied BIM techniques, such as identification of data requirements, operational methods and contractual processes; and
- driving domestic and international growth in technology and construction, including boosting employment opportunities.

As global data traffic is set to quadruple by 2016 and around 60 per cent of digital data will be cloud based, the digitisation of construction is perceived by Government to be a prime mover in driving that process of change. In time, and from a retrospective viewpoint, the 2016 threshold for the use of BIM on central government-procured projects may prove to have been a significant milestone along the pathway towards realising 2025 construction targets. On the other hand, it may well be the case that the Government Gateway did not significantly influence the way that UK construction operates. It could be argued that whether or not the Government had mandated on BIM, large organisations with global presence – for example, international BIM-smart companies such as Ramboll, with almost 300 offices in 35 countries – would probably have pushed ahead with digitisation anyway and carried smaller supply chain partners in the wake of their forward travel.

Looking back at previous industry initiatives, such as the Latham (1994) and Egan (1998) Reports, as Andrew Wolstenholme reported in 2009, in the 11 years after the Egan Report was published, there had been some progress, but not nearly enough. Few of the Egan targets had been met in full, while most had fallen considerably short. Where improvement had been realised, too often the commitment to Egan's principles appeared skin-deep. In some sectors, such as housing, embodying innovation into operational methodologies could be inhibited by limited understanding of how added value can be created through making the construction process more efficient.

In reviewing inhibitors to change, Wolstenholme's team also concluded that few construction organisations have the purchasing power to leverage their supply chains, or the resources to invest heavily in IT. Those which did have the purchasing power failed to maximise their leverage, and only a handful of construction firms offered a vertically integrated approach from design to managed handover. With that kind of track record, it may seem a tall order to expect the UK's SMEs and micros to aspire to Level 3 BIM when, at the mid point of 2015, there is little hard evidence of widespread engagement with Level 2 conventions and practice.

The fragmentation of the UK construction 'industry' has often been cited as a barrier to establishing unanimity of purpose when it comes to implementing change. In that context, the authors have discussed how the UK construction scene is characterised by a proliferation of SMEs and micro organisations. The built environment design, construct and maintain supply chain includes designers, surveyors, contractors, manufacturers, suppliers and specialists. Sometimes these individuals and organisations work independently in servicing client requirements on a project-by-project basis. On other occasions, small firms can make trusting and sustained informal business relationships: organisational clusters, organic in nature, agile and metamorphic in their ability to sustain and adapt over time.

A great deal has been written about BIM; maybe too much for some people, particularly those from SMEs and micro organisations. Anecdotal evidence suggests that some smaller firms feel excluded by industry promotions which give the impression that BIM is applicable to large companies and very large projects. That perception needs to be challenged. MacLeamy has argued that BIM is quintessentially about how we collaborate to use digital data efficiently to drive the processes involved in making and using buildings. That premise can be applied universally, whatever the project size.

A 2013 study carried out for the UK Government by EC Harris suggested that for a building project in the £20–25 million range, the main contractor may be directly managing around 70 subcontracts, of which a large proportion may be small (typically £50,000 or less). Smaller projects demonstrated up to 70 per cent of subcontracts having a value of £10,000 or less. Clearly, even with large publicly procured projects, because of the fragmented and complex UK supply chain morphologies, change can permeate down and touch even the smallest organisations.

SMEs and micros need to be engaged and aware that they may already be working to BIM Levels 0 and 1 anyway, so that it might not be a massive leap for them to attain Level 2 capability. As with any paradigm shift which is likely to impact on work practices, sometimes it is difficult to filter out propaganda from evidence-based observation and recording of the cause and effect of change. The way in which the UK BIM-scape has started to unfold over the last couple of years might offer a case in point to support that argument.

To what extent the UK Government's BIM strategy will impact on the supply chain's behaviour, and in particular leverage change among the UK's SMEs and micros remains to be seen. As Peter Trebilcock (2014) noted:

> no one seems to be asking for evidence of how BIM is improving safety, how it dovetails into off-site construction, how it saves estimating time or how it is helping to eradicate rework. Many can demonstrate these advantages through impressive stats and case studies, but does anyone care?

The literature does suggest that (with some exceptions) large organisations are leading the charge. At this point in time that may well be the case, particularly if signals given out by the so-called early adopters are a reliable indicator. There is almost a Darwinian inevitability in the ability of larger firms to carry the overheads of BIM trouble shooters and the on-costs of engaging with unwieldy prequalification documentation as part of the process of winning work, particularly public sector projects. Equally, the observation applies to the evolution of BIM standards, which have put down markers as normative reference points for UK industry.

In terms of the Government's aspirations, BIM UK has now been effectively ring-fenced by a series of reference documents. The cornerstone for these benchmarks has been provided by the PAS 1192 suite of standards. The genealogy of the PAS 1192-2 template can be traced back to the watershed Heathrow T5 project. As a paradigm, that scenario presented a very particular set of circumstances. Do these necessarily represent the full breadth of our industry, and does the anchor PAS 1192-2 document extrapolate comfortably across a wide spectrum of project types and values? Certainly, the follow-up PAS 1192-3 seems to have lost the spirit of the soft landings concept developed in the late 1990s by architect Mark Wray and championed since then by BSRIA. With a focus on the operational and fiscal aspects of 'asset management', the building user's perspective seems to have been redacted from the script. Similarly, even small built environment organisations with an empathy towards using BIM techniques may struggle to find the PAS 1192-4 COBie protocol fitting comfortably with their practice.

Clearly, built environment is a giver, in that the sector makes a huge contribution to the well-being of the UK economy. But the flip side, as Ray Crotty has noted, is that construction is also a significant taker. Buildings are responsible for almost 50 per cent of the UK's carbon emissions, 50 per cent of our water consumption, 30 per cent of landfill waste and around 25 per cent of all raw materials used in the economy. Across mining, quarrying, construction and demolition, the construction sector accounts for just under two-thirds of the waste generated in the UK. Can BIM do anything significant to ameliorate this situation? With SMEs and micros representing up to 90 per cent of the UK construction community by numbers of firms, the extent to which small organisations engage with digitisation of the industry may well be a significant factor in determining whether the UK Government's ambitious 2025 built environment targets can be realised.

Bibliography

Anderson, C. (2012) *Makers: The New Industrial Revolution*. London: Random House Business Books.

Anon (2013) 'Government Soft Landings (GSL): An Overview', HM Government, [online]. www.bimtaskgroup.org/wp-content/uploads/2013/02/GSL-Overview-for-Web-site-01-03-13-ver-a-Read-Only.pdf [accessed 19 January 2015].

Anon (2014) 'BIM Key to Eliminating Building Data Risk', FM World, 1 April, [online]. www.fm-world.co.uk/news/fm-industry-news/bim-key-to-eliminating-building-data-risk/ [accessed 19 January 2015].

Anon (2015a) *Digital Built Britain, Level 3 Building Information Modelling – Strategic Plan*, HM Government, [online]. http://digital-built-britain.com/DigitalBuiltBritainLevel3BuildingInformationModelling StrategicPlan.pdf [accessed 24 March 2015].

Anon (2015b) 'Skills Shortage: Unis Struggle to Recruit to Degrees in Construction', *Construction Manager*, 27 January, [online]. www.construction-manager.co.uk/news/skills-shortage-special-u nis-struggling-recruit-bu/ [accessed 27 January 2015].

Baba, A., Mahjoubi, L., Olomoilaiye, P. and Booth, C. (2013) 'Decision Support Framework for Low Impact Housing Design in the UK', *Proceedings 29th Annual ARCOM Conference*, Reading, UK, [online]. www.arcom.ac.uk/-docs/proceedings/ar2013-1341-1350_Baba_Mahjoubi_Olomoilaiye_Colin.pdf [accessed 14 January 2015].

Brand, S. (1994) *How Buildings Learn*. New York: Penguin Books.

Cohen, J. (2010) *Integrated Project Delivery Case Studies*, AIA California Council, [online]. www.aia.org/aiaucmp/groups/aia/documents/pdf/aiab082051.pdf [accessed 19 January 2015].

E4BE (2013) 'BIM: Higher Education and Growth', report of workshop held at Construction Industry Council, London, 5 November, [online]. http://cic.org.uk/networks-and-committees/e4begroup.php [accessed 19 October 2014].

Grabher, G. (2004) 'Learning in Projects, Remembering in Networks? Communality, Sociality and Connectivity in Project Ecologies', *European Urban and Regional Studies*, 11(2), [online]. www.lse.ac.uk/geographyAndEnvironment/pdf/grabher_1203.pdf [accessed 19 January 2015].

Greenfield, S. (2014) *Mind Change: How Digital Technologies Are Leaving their Mark on our Brains*. London: Rider Books.

Griffith, E. (2013) 'What is Cloud Computing?', *PC Magazine*, 13 March, [online]. http://uk.pcmag.com/networking-communications-software-products/16824/feature/what-is-cloud-computing [accessed 19 January 2015].

Jack, M. (2014) *Housing Innovation Showcase 2012, Building Performance Evaluation Phase 1 – Part 1*, Kingdom Housing Association, [online]. www.housinginnovationshowcase.co.uk/media/3e8205c5df b26350ffff8111ffffe417.pdf [accessed 14 January 2015].

Khemlani, L. (2011) 'BIM for Facilities Management', *AECbytes*, [online]. www.aecbytes.com/feature/2011/BIMforFM.html [accessed 14 January 2015].

Knutt, E. (2015) 'BIM: Is the Industry Ready?', *CIOB BIM+*, Chartered Institute of Building, 7 January, [online]. http://bim.construction-manager.co.uk/news/bim-industry-ready/ [accessed 19 January 2015].

Leaman, A. and Bordass, B. (2010) 'Assessing Building Performance in Use 4: The Probe Occupant Surveys and their Implications', *Building Research & Information*, 29(2), 2001, [online]. www.tandfonline.com/doi/abs/10.1080/09613210010008045# [accessed 26 March 2015].

Limburg, D. (2012) 'Ready, Willing and Capable: How Can SMEs Gain Advantage from Using Internet-based Technologies', *Proceedings of the 17th Annual UK Academy for Information Systems Conference*.

UKAIS, Oxford, [online]. www.ukais.org.uk/documents/downloads/conference201279ecac83-a d40-4713-a084-e1e8bb4d658d.pdf [accessed 19 January 2015].

Macdonald, J.A. (2012) 'A Framework for Collaborative BIM Education across the AEC Disciplines', Proceedings of *37th Annual Conference of Australasian University Building Educators Association (AUBEA)*, 4 6 July, Sydney, Australia, [online]. https://www.academia.edu/1767377/A_ Framework_for_Collaborative_BIM_Education_Across_the_AEC_Disciplines   [accessed   27 January 2015].

Macdonald, S., Anderson, P. and Kimbel, D. (2000) 'Measurement or Management?: Revisiting the Productivity Paradox of Information Technology', *Vierteljahrshefte zur Wirtschaftsforschung* (Quarterly Journal of Economic Research), 69(4), [online]. www.diw.de/documents/publikationen/73/3873 9/v_00_4_9.382949.pdf [accessed 19 January 2015].

Martin, C., Lowson, R. and Peck, H. (2004) 'Creating Agile Supply Chains in the Fashion Industry', *International Journal of Retail and Distribution Management*, 32(8), [online]. https://dspace.lib.cranfield. ac.uk/bitstream/1826/2651/1/Creating%20agile%20supply%20chains-fashion%20industry-2004. pdf [accessed 14 January 2015].

Miller, D. (2012) 'BIM from the Point of View of a Small Practice', Building Information Modelling, NBS,   [online].   www.thenbs.com/topics/bim/articles/bimsmallpractice.asp   [accessed   19 January 2015].

OECD (1996) *The Knowledge Based Economy*, Organisation for Economic Co-operation and Development, Paris, [online]. www.oecd.org/sti/sci-tech/1913021.pdf [accessed 19 January 2015].

Pringle, J. (2012) 'Fragmented Construction Industry', *Building*, 5 October, [online]. www.building. co.uk/fragmented-construction-industry/5043358.article [accessed 19 January 2015].

Rosenfield, K. (2012) 'The Future of the Building Industry: BIM-BAM-BOOM!', *ArchDaily*, 9 August, [online]. www.archdaily.com/262008/the-future-of-the-building-industry-bim-bam-boom/ [accessed 14 January 2015].

Scott, M. (2014) 'Better Together: Why Construction Needs Collaboration to Work Efficiently', *The   Guardian*,   17   July,   [online].   www.theguardian.com/sustainable-business/collaboration-construction-buildings [accessed 19 January 2015].

Trebilcock, P. (2014) 'Who is Measuring BIM?', *Building*, 3 November, [online]. www.building.co.uk/ who-is-measuring-bim?/5071869.article [accessed 14 January 2015].

Ward, S. (2015) '5 Disadvantages of Cloud Computing: Consider These Before You Put Your Small Business in the Cloud', *About Money*, [online]. http://sbinfocanada.about.com/od/itmanagement/a/ Cloud-Computing-Disadvantages.htm [accessed 20 January 2015].

Webb, B. and Schlemmer, F. (2008) *Information Technology and Competitive Advantage in Small Firms*. London: Routledge Studies in Small Business.

Wolstenholme, A. (2009) 'Never Waste a Good Crisis', *Constructing Excellence*, 30 November, [online]. http://constructingexcellence.org.uk/wp-content/uploads/2014/12/Wolstenholme_Report_ Oct_20091.pdf [accessed 24 March 2015].

BIM glossary

3D	Three-dimensional geometry
4D	Construction sequencing information
5D	Cost information
6D	Project life cycle information
Actual Digital Questions (ADQ)	These break down Plain Language Questions (PLQ) in a way that means they can be answered with a digital answer
Assembly	A physical aggregation of components
Asset Information Model (AIM)	A model that compiles the data and information necessary to support asset management
Asset Information Requirements (AIR)	Define the information required for an Asset Information Model
Attribute	A piece of data that describes a characteristic of an object or entity
BIM Collaboration Format (BCF)	An open file format based on XML that allows the addition of comments to an Industry Foundation Classes BIM model
BIM Execution Plan (BEP)	A developing strategy prepared by suppliers that comprises a pre-contract BIM execution and then a post-contract BIM execution plan
BIM maturity levels	The levels of complexity and collaboration that building information modelling can take
BIM protocol	A supplementary legal agreement that can be incorporated into professional services appointments, construction contracts, subcontracts and novation agreements
BIM Task Group	Bringing together expertise to strengthen the public sector's BIMcapability and provide the information the industry needs to meet the government's BIM requirement
BS1192:2007	Collaborative production of architectural, engineering and construction information – code of practice
BS1192-4:2014	Collaborative production of information Part 4: Fulfilling employer's information exchange requirements using COBie – code of practice
BS8541	Library objects for architecture, engineering and construction
BS ISO 16739	Industry Foundation Classes (IFC) for data sharing in the construction and facility management industries
BS ISO 55000	Asset management – overview, principles and terminology
Building Information Modelling (BIM)	A very broad term that describes the process of creating a digital model of a building or other asset (such as a bridge, highway, tunnel and so on) using object-oriented information
CAD	Computer-Aided Design
CAFM	Computer Aided Facility Management

CAPEX	Capital Expenditure
CIC BIM Protocol	The Construction Industry Council Building Information Modelling Protocol: standard protocol for use in projects using Building Information Models
Clash Rendition (CR)	A rendition of a building information model specifically to avoid clashes in spatial co-ordination
CMMS	Computerised Maintenance Management System
Common Data Environment (CDE)	The single source of information for the project, used to collect, manage and disseminate documentation, the graphical model and non-graphical data for the whole project team
Component	A physical item or feature
Construction Operations Building Information Exchange (COBie)	A non-proprietary multi-page, spreadsheet data format for the publication of a subset of building information models focused on delivering asset data rather than geometric information
Contractor's proposals	Set out contractor's proposals for designing and constructing a built asset, along with their price
CPI	Construction Project Information
CPIc	Construction Project Information Committee
CPIx	Construction Project Information Xchange
CSG	Constructive Solid Geometry
Data	Information that has not been interpreted
Data drops	Information of a particular format and level of detail issued to the client at predefined stages of a project. Also known as information exchange.
Data manager	A procedural gate-keeper who sets up and manages an Asset Information Model
Design intent model	The initial version of the Project Information Model (PIM)
DGN	MicroStation and Intergraph file format
Digital Built Britain	Digital Built Britain, Level 3 Building Information Modelling – Strategic Plan
DWF	AutoCAD Design Web Format
DWG	AutoCAD native file format
DXF	Interchange file format
Employer's Information Requirements (EIR)	Set out the information required by the employer aligned to key decision points or project stages, enabling suppliers to produce an initial BIM execution plan from which their proposed approach, capability and capacity can be evaluated
Employer's requirements	Provide a description of all the client's requirements for a built asset, in response to which contractors prepare contractor's proposals
Employer's decision point	Project gateway aligned to key stages at which information is provided to answer the employer's plain language questions and a decision is taken about whether to proceed
Federated building information model	An assembly of distinct models to create a single, complete building information model of an asset
Government Soft Landings (GSL)	The process of aligning the interests of those who design and construct an asset with the interests of those who use and manage it. GSL sits alongside BIM, as BIM feeds into Computer Aided Facility Management (CAFM) systems, and helps enable future alterations to completed buildings.
iBIM	Integrated Building Information Model, or Level 3 BIM
ICT	Information and communications technology
IFC2x	Industry Foundation Class version 2x
Industry Foundation Classes (IFC)	A neutral, non-proprietary data format used to describe, exchange and share information
Information Delivery Manual (IDM)	Specifies when certain types of information are required during the construction of a project or the operation of a built asset. Now known as the buildingSMART® standard for processes.

Information exchange	Information of a particular format and level of detail issued to the client at predefined stages of a project. Also known as data drop.
Information management process (IMP)	Procedures implemented to manage the Asset Information Model
Information manager	A procedural gate-keeper who sets up and manages the Common Data Environment (CDE)
International Framework for Dictionaries (IFD)	A standard for terminology libraries or ontologies. It establishes specific obligations, liabilities and limitations on the use of building information models and can be used by clients to mandate particular working practices.
Layer	Attribute that allows control over the visibility of entities within CAD files
Level 0 BIM	Unmanaged computer aided design (CAD) including 2D drawings, and text with paper based or electronic exchange of information but without common standards and processes
Level 1 BIM	Managed CAD, with the increasing introduction of spatial co-ordination, standardised structures and formats as it moves towards Level 2 BIM
Level 2 BIM	Managed 3D environment with data attached, but created in separate discipline based models
Level 3 BIM	A single collaborative, online, project model with construction sequencing (4D), cost (5D) and project life cycle information (6D). This is sometimes referred to as 'iBIM' (integrated BIM).
Level of definition	Overall term describing the level of model detail and the level of information detail
Level of model detail (LOD)	The level of detail of graphical content in building information models
Level of model information (LOI)	The level of detail of non-graphical content in building information models
Master Information Delivery Plan (MIDP)	The primary plan for the preparation of the project information required by the Employer's Information Requirements, setting out programme, responsibility, protocols and procedures
Master Information Document Index (MIDI)	A detailed list of information deliverables for a project identifying responsibility for delivery and programme
Model Production and Delivery Table	Referred to in CIC BIM protocol, summarising originators of and level of detail for the Project Information Model at different stages of the project
OPEX	Operating Expenditure
Organisational Information Requirements (OIR)	The information required at an organisational level (rather than asset level) to achieve the business objectives of the organisation
Origin	The setting-out point for a project
Parametric modelling	The generation of a model based on a series of pre-programmed rules or algorithms
PAS 1192-2:2013	Specification for information management for the capital/delivery phase of construction projects using building information modelling
PAS 1192-3:2014	Specification for information management for the operational phase of construction projects using building information modelling
Plain Language Questions (PLQ)	Questions that the client will want to ask at key stages of the project to decide whether the project has developed satisfactorily and whether it should proceed
Post-contract BIM execution plan	Confirms the supply chain's capability, capacity and competence to meet the Employer's Information Requirements and providing a Master Information Delivery Plan
Pre-contract BIM execution plan	Prepared by prospective suppliers, setting out their proposed approach, capability, capacity and competence to meet the Employer's Information Requirements

Process map	Identifies the key decisions that will need to be made during the project to ensure that the solution developed satisfies the business need that has been identified
Project execution plan (PEP)	Sets out the overall strategy for managing a project, describing the policies, procedures and priorities that will be adopted
Project Implementation Plan (PIP)	A statement relating to the suppliers' IT and human resources capability to deliver the Employer's Information Requirements submitted as part of the pre-contract BIM Execution Plan by each organisation bidding for a project
Project Information Model (PIM)	The building information model developed during the design and construction phase of a project
RACI indicator	Identifies authorities of participants in relation to specific project activities. Responsible, Authorise, Contribute, Informed.
RAG report	Report giving status assessments in terms of Red, Amber or Green
Standard Method and Procedure (SMP)	Defines how information is named, expressed and referenced
Supplier information modelling assessment form	Sets out the capability and competence of a supplier to use BIM in a collaborative environment
Supplier information technology assessment form	Sets out the capability and IT resources of a supplier for exchanging information in a collaborative environment
Supplier resource assessment form	Used to assess a supplier's resource capability and capacity, based on the supplier information technology assessment form and supplier information modelling assessment form
Supply chain capability assessment form	Summarises the human resource and IT capability of each suppliers that make-up the supply chain
Supply Chain Capability Summary (SCCS)	Prepared by the principal supplier to verify that there is suitable human resource and IT capability in the various organisations that make up the supply chain
System	Set of components with a common function
Task Information Delivery Plan (TIDP)	Prepared by individual task team managers, setting out the responsibilities for each individual information deliverable
Third party capability assessment form	Sets out the information management and IT capability of non-design, non-construction suppliers. This is made up of individual Task Information Delivery Plans (TIDP).
Virtual Construction Model	Development of the design intent Project Information Model
Volume	A manageable spatial subdivision of a project that allows more than one person to work on the project models simultaneously
xBIM (eXtensibleBuilding Information Modelling)	An open-source software development tool that allows developers to read, create and view BIM in the IFC format
XML	Extensible Mark-up Language, for encoding data in a format that is human-readable and machine-readable
Zone	A set of locations with a shared attribute

Index

Milton Keynes UK
Ingram Content Group UK Ltd.
UKHW051537141024
449569UK00028B/1508